THE MAYAN CODE

"Barbara Hand Clow's inspired and well-researched articulation of the Mayan Calendar provides a significant aid to understanding the mechanics of the acceleration of time and the process of human evolution. She offers a refreshingly radical and hopeful outcome for our future beyond the Calendar's end in 2011–2012."

Nicki Scully, author of *Alchemical Healing* and coauthor of *Shamanic Mysteries of Egypt*

"From now through 2012 we are walking what the Mayas call *Xi Balba bih,* 'The Road of Awe.' Tools that assist us in looking at the vast acceleration and wave of evolution that we are experiencing, through the largest possible soul lens, will be of great assistance during these transformative times. This book is highly recommended for that process."

Ariel Spilsbury, coauthor of *The Mayan Oracle: Return Path to the Stars*

THE MAYAN CODE

Time Acceleration and Awakening the World Mind

Barbara Hand Clow

Illustrations by
Christopher Cudahy Clow

Bear & Company
Rochester, Vermont

Bear & Company
One Park Street
Rochester, Vermont 05767
www.BearandCompanyBooks.com

Bear & Company is a division of Inner Traditions International

Library of Congress Cataloging-in-Publication Data

Clow, Barbara Hand, 1943–
The Mayan code : time acceleration and awakening the world mind / Barbara Hand Clow ; illustrations by Christopher Cudahy Clow.
p. cm.
Summary: "Bestselling author Barbara Hand Clow shows how the Mayan Calendar is a bridge to galactic wisdom that fosters personal growth and human evolution"—Provided by publisher.
Includes bibliographical references and index.
ISBN-13: 978-1-59143-070-4
ISBN-10: 1-59143-070-4
1. Maya calendar. 2. Maya astrology. 3. Maya cosmology. I. Title.
F1435.3.C14C56 2007
529'.32978427—dc22

2006103125

Printed and bound in the United States by Lake Book Manufacturing

10 9 8 7 6 5 4 3 2

Text design by Rachel Goldenberg
Text layout by Virginia Scott Bowman
This book was typeset in Sabon with Trajan as the display typeface

All of the illustrations in this book are by Christopher Cudahy Clow, unless otherwise noted.

To send correspondence to the author of this book, mail a first-class letter to the author c/o Inner Traditions • Bear & Company, One Park Street, Rochester, VT 05767, and we will forward the communication.

This book is dedicated to my oldest son,
Tom, who died in June of 2004.

Tom, you had such a fine mind. Since you couldn't
tell this story in this dimension—yet you transmitted
it through me—I have
written it for you.

I will always wish I had some time with you sitting
by the Inukshuk on the edge of
English Bay in Vancouver, BC.

CONTENTS

LIST OF ILLUSTRATIONS

FOREWORD

In the legends of the ancient Maya, a divine incarnation who went by the name First Mother played a significant role. According to the Temple of the Inscriptions in Palenque, this divinity was born on December 7, 3121 BC, to prepare for the creation of what we may now call the National Underworld (the original Maya name is unknown). Surprisingly, perhaps, although it happened first, the birth of First Father, June 16, 3122 BC (exactly seven years before the onset of the National Underworld through his activation of the World Tree), was mentioned only *after* that of the First Mother. This attests to the importance of the First Mother, who was later also to take a pivotal part in the dynastic politics of Palenque, where the shaman King Pacal is known to have manipulated his own birthday to prove his spiritual heritage from her.

Barbara Hand Clow, "Grandmother Sky," has literally played a role in the modern world corresponding to that of First Mother in the retrieval of the Mayan calendrical knowledge. Thus, through her publishing activities at Bear & Company, she helped disseminate to the world the books that pioneered this genre. Not only did she publish the early works by Arguelles and Jenkins on the Mayan Calendar, her advice as a reviewer was instrumental in the later publication of one of my own books by Bear & Company. Throughout the time she was active as a publisher, she herself wrote a series of books, among which I find especially relevant *The Mind Chronicles* and *The Pleiadian Agenda*. She has also taught countless students, who have come to experience many aspects of Native American and Maya knowledge through her work. Like the Maya, in the Pleiades she sees an origin for her information.

Since she has seen this field be born and grow to its current extent, it is at present very valuable to read the synthesis of her thoughts in *The Mayan Code,* not only of the current state of the art of the study of the Mayan Calendar, but also of her extensive knowledge about astrology, esotericism, and shamanism.

What is meant by "at present"? As always, when it comes to studies related to the Mayan Calendar, it is critical to observe when a phenomenon emerges and what energy brings it into the world. The whole point of the Mayan Calendar is that it is not a linear calendar counting astronomical cycles, but one that describes the shifting spiritual energies of time. In such an energetic perspective, *The Mayan Code* comes onto the shelves shortly after the Fifth Day of the Galactic Underworld began on November 24, 2006. Today it is critical that information about the prophetic meaning become broadly accessible in a rigorous and yet attractive form to the public at large. At a time when Mel Gibson's film *Apocalypto* has raised mainstream awareness of the existence of the Mayan Calendar, it is vital that we move beyond the question "What will happen in the year 2012?" to the deeper question of what is happening at the current time and what divine creation is all about. I believe we can understand what will happen in the year 2012 only if we realize that we are living in an ongoing divine process of creation and its timelines. So far, only a small minority of humanity is aware of this.

Because its mission is to strengthen the right brain, the Fifth Day of the Galactic Underworld will see the creation of new syntheses, especially of science and spirituality. The Fifth Day is also ruled by the energy of Quetzalcoatl or Christ, which in any particular Underworld leads to a decisive breakthrough of its particular phenomena. Since the Galactic Underworld fundamentally brings an end to Western dominance in the external world, and an integration of the intuitive and rational aspects of self in the internal world of the individual, *The Mayan Code* comes at a time when it is critically needed. It is crucial for people to understand the altered relationships in the world that this Underworld will bring as they are playing out in this very moment. In this regard, this book will play a significant guiding role. Now is the time for the Mayan Calendar to be studied and debated; when we come to the year 2012, it will surely be too late.

By 2012, we will either have made it through and been able to fulfill the divine plan, or have failed to do so because of our ignorance or the disinformation being spread by particularist interests. Regardless, in 2012 our ability to influence the course of events toward a collective quantum jump in consciousness to the Universal Underworld will have been lost. This is why we should focus on what the Mayan Calendar is telling us about the present moment and not about 2012. The serious student of the Mayan Calendar is not sitting around waiting to see what will happen, but is seeking to contribute to the ability of everyone to make this quantum jump. It deserves to be pointed out here that much of the debates over calendrical features and end-dates that have been going on over the years are not just hairsplitting. The positions taken on such key issues, which not too long ago seemed to be of interest only to scholars in dusty libraries, are now turning out to have decisive consequences for how we relate to humanity's future.

Far from these dusty libraries, I am convinced that the Mayan Calendar has now become the Theory of Everything (or complete theory) that was anticipated in the closing words of Stephen Hawking in *A Brief History of Time:*

> If, however, we were to discover such a complete theory, its general principles should be so simple as to be understood by everyone, not just a few scientists. Then we shall all, philosophers, scientists, and just ordinary people, be able to take part in the discussion of the question of why it is that we and the universe exist. If we find the answer to that, it would be the ultimate triumph of human reason—for then we would know the mind of God.

The book you are holding in your hand is a significant contribution to the abovementioned discussion. Hence, while the Mayan Calendar system lays down the universal framework for us to understand the one truth about God and God's creation, we must recognize that no single human being is able to grasp the fullness of this truth. Our perspectives are always limited by our particular backgrounds and the fact that we are very much a part of the world we want to understand. Because of these limitations, there is a need for a discussion between different individual perspectives. This book by Barbara Hand Clow, the First Mother of modern Mayanism, may be regarded as an invitation to all to participate in the broader discussion of why we and the universe exist. Only as modern human beings have been able to integrate the contribution of the ancient Western Native American civilizations has it become possible for our worldview once again to become whole. While many recognize the value of the reverence that Native Americans have had toward nature, it is long overdue for us to realize that without their intellectual contribution, the calendar system, we would never have arrived at the framework of a complete theory. The Mayan Calendar has been the missing piece of the puzzle, and fortunately more and more people are becoming aware of this.

The Mayan Code brings new perspectives on all the Underworlds that constitute the prophetic Mayan Calendar system. The liveliness and tangibility of Barbara Hand Clow's description of time acceleration, which is another word for the divine process of creation, is unsurpassed. I am certain that the reader will truly enjoy being on a journey with her as she explores its consequences for our own lives and for the universe at large. Her writings lead us to contemplate many unsolved mysteries, ranging from Neolithic stone settings to intragalactic communications and the nine dimensions of consciousness. The students of mine who have been dissatisfied with the absence in my theories of any reference to Atlantis and lost civilizations will here also find an alternative view.

I find especially interesting her explanations as to why it is that the Galactic Underworld, at least so far, has appeared so "ungalactic." It has seemed to be much more about the movement toward balance on our own planet and in our own minds than about expanding into the Galaxy. It is in the very nature of things, however, that the totally new phenomena of the future are unpredictable. Nobody was able to predict the European discovery of the Americas, or the emergence of the Internet, for instance, and it always seems more difficult to predict completely new phenomena that expand and enhance our world than catastrophes that may destroy it. We must consider it likely that completely unpredictable phenomena will enrich our world, and maybe, if Barbara Hand Clow is right, we are in for a surprise.

The famous Austrian philosopher Ludwig Wittgenstein once said that his theses were like a ladder that was absolutely necessary for someone who wanted to climb to the roof to have an overview of things, but once the roof had been attained, that ladder would no longer be needed and could be thrown away. The Mayan Calendar is similar. Barbara Hand Clow sums it up well: "During 2012, all the seasonal festivals, the equinoxes and the solstices, will be celebrated. And when time ends and the evolutionary activation driven by the World Tree is finally complete, the people of Earth will have forgotten all about history and the Mayan Calendar; they will be in ecstatic communion with nature and the Creator."

CARL JOHAN CALLEMAN
BELLINGHAM, WASHINGTON
10 CABAN, 7.16.17 OF THE GALACTIC UNDERWORLD
(OCTOBER 31, 2006)

Carl Johan Calleman holds a Ph.D. in physical biology and has served as an expert on cancer for the World Health Organization. He began his studies on the Mayan calendar in 1979 and now lectures throughout the world. He is the author of *The Mayan Calendar and the Transformation of Consciousness* and *Solving the Greatest Mystery of Our Time: The Mayan Calendar.*

ACKNOWLEDGMENTS

I want to thank Carl Johan Calleman for discovering the time-acceleration factor in the Mayan Calendar, and for his unshakable devotion in bringing this news to the public. Carl is a devoted researcher and friend who always had time to spare when I hammered him with questions and piles of paper. Then to top that, he even took the time to write a wonderful foreword when I finished the book! You, Carl, are changing the world—possibly more than you realize.

As always, working with my son Christopher was a joy. Being a fellow artist with one's child is possibly the greatest happiness a mother can experience. When I was young and enjoying the works of R. A. Schwaller de Lubicz and his wife, Isha, I loved the way they worked with their daughter, the illustrator Lucy Lamy. I always hoped I'd have a child who would work with me in the same way, and I do! Something special happens, Chris, when you use your inner artist's eye to make my thoughts visible, so thank you!

Richard Drachenberg's editorial assistance greatly improved this text, but Richard, your work meant much more to me than that. Writing this book was daunting because I had so little time, and I had to hurry way too much due to the fact that the material contained herein is of such extreme importance. I felt the urgency of my task, and often, when I doubted my ability to get through it, your encouragement kept me going. I thank you very much for your profound contributions.

I thank Ian Lungold for writing the Conversion Codex for the Mayan Calendar (appendix D), and I thank Ian's partner, Matty, for giving me generous permission to use it. It is the best codex I have ever found because it is so easy to use! And thank you, Gerry Clow, for editing the Codex and making it just perfect. Thanks for all the other help you gave on this book, especially editing the final page proofs. As always, your support and assistance has been given with love and concern for the readers. And thank you, Louisa McCuskey, for helping on the artwork for the Conversion Codex. You have done a beautiful job.

As for Inner Traditions/Bear & Company, I am so grateful to Jon Graham for seeing the importance of this book. It was your encouragement, Jon, that helped me realize I must write it. Thank you, Ehud Sperling, for being at the helm of one of the greatest publishing houses ever. Thank you to my copy editor Judy Stein and thank you to Jeanie Levitan, Anne Dillon, Peri Champine, Rob Meadows, and all the other great people at Inner Traditions/Bear & Company.

INTRODUCTION

Since 1987, three great modern philosophers have published masterworks on the Mayan Calendar: José Argüelles with *The Mayan Factor* in 1987; John Major Jenkins with *Maya Cosmogenesis 2012* in 1998; and Carl Johan Calleman with two versions of *The Mayan Calendar* in 2001 and 2004. I have written many books myself since 1986 that occasionally touch on the Mayan Calendar, and now I'm ready to enter into this discussion just when the end of the Calendar is only five short years away.

You've probably heard that the conquering Spaniards burned up most of the Maya, Nahuatl, and Aztec literature. You may think there is not much left to investigate about the Maya, except possibly to visit their fantastic pyramid sites. In fact, in Maya studies, we have as much mythology, literature, and art as we have from the ancient Egyptians, Sumerians, or Greeks. There are still many descendants of the Maya who hold ancient memories very much alive, especially by keeping the ancient calendars. Maya culture is a gold mine. In this book, I will investigate the Maya science of time cycles, primarily by means of the Mayan Calendar, which tracks the last 5,125 years of history, the Long Count, which is also called the "Great Year" by Calleman.

What special qualifications do I have that might offer any more wisdom in this maturing investigation? From 1982 until 2000 I was the co-publisher of Bear & Company, then located in Santa Fe, New Mexico. We published the work of José Argüelles, John Major Jenkins, and other Maya researchers. Then in 2004, Bear published Carl Johan Calleman. As things turned out, by living and working in this field during the 1980s through early 2000, I was exposed to a whole new vista of what the Maya thought about more than a thousand years ago. This vista is what I call the Maya time-acceleration legacy, a view that is very important for the whole world now.

Ultimately, my understanding of the Calendar is very intuitive, since I am an active indigenous and Western shaman trained to travel in many worlds. Maya shamans initiated me at many of their sacred sites while I studied the Calendar. It is time to share what I learned from them, which I know my Maya teachers, Hunbatz Men and Don Alejandro Oxlaj, and my Cherokee teachers, my grandfather Gilbert Hand and J. T. Garrett, will appreciate. My understanding of how the Pleiades star system is part of Cherokee/Maya thought is especially valuable, because since birth I have been living simultaneously on Earth and in the mind of Alcyone, the central star of the Pleiades. In my childhood, Grandfather Hand taught me that Earth and Alcyone are my homes. I've never forgotten my star origins—not for a moment—since the Maya and the Cherokee are the people of the Pleiades.

All this having been said, there is something more deeply personal that guided the writing of this book. Like José Argüelles, who lost his oldest son, Josh, a week after Harmonic Convergence in 1987, I have lost two of my own sons during critical turning points in the Calendar. My sons Tom and Matthew both participated with me in Maya initiatic training (which is described in a moment), and I feel that their energy from the other side is very much a part of this book. I hope the knowledge they have transmitted to me from the realms of spirit has deepened my work—at the same time, this experience with them has helped me through many stages of grieving.

You've probably heard that nothing is more traumatic and challenging than losing your own child, and this is absolutely true. Yet, as with all other kinds of losses, if you can learn from such deep pain and transcend it, you are capable of attaining an overview of what's truly important about life. The indigenous people of Mexico and Guatemala have suffered genocide many times over, which has almost destroyed their cultures, and still they are waiting for us to listen to them. I have survived my own personal wipeout of half of my family, and the only good thing I have gotten out of it is that I understand the suffering people of our planet. As families suffer daily in Iraq, I experience deep feelings of compassion for their losses. Before saying a few things about losing two of my sons, I ask every mother and father to resist having their sons and daughters sent to war by old men who do not care if these children die.

If they die there, you will know it was not worth it.

In their respective works on the Mayan Calendar, Argüelles, Jenkins, and Calleman each conclude that Earth will not survive if the Western scientific, materialistic, and progress-oriented paradigm continues to consume the world. All three teach that learning the secrets of the Maya can inspire humans to become enlightened and derail the Western imperialistic train to hell. As you will see in this book, I believe we *will* derail that train to hell, and the way we will do it is to create peace in our world.

My oldest sons Tom and Matthew were young men with sensitive hearts; both were philosophers and ecologists. My sons' deaths were intimately involved in my awakening to the wisdom of the Calendar, so we need to know a little more about Tom and Matthew.

As you will see in this book, Earth experienced an alignment to the center of the Milky Way Galaxy in 1998, which has fundamentally altered the physical and psychic field of our planet; this is a scientific fact. This occurred while I was in Bali doing rainmaking ceremonies to extinguish the fires in the Kalimantan forests of Borneo. The rains came, and the Balinese invited me into the inner recesses of their most sacred temples to celebrate. I was in a state of ecstasy when I came home, and I now realize that my shamanic experience in Bali coincided with Earth's alignment to the Milky Way. I shared these amazing experiences on the phone with my twenty-nine-year-old son, Matthew.

He'd never read my work because he believed that reading my books would scramble his ability to think logically, which could have rendered him useless during his graduate studies in limnology (lake studies). Oddly enough, though, during our last conversation, he said that he had just read my 1995 book, *The Pleiadian Agenda*. He said that my work was as important for the planet as his own work as a budding ecologist, which meant more to me than anything anyone has ever said about my writing. One month after our phone conversation, Matthew drowned in Red Rock Lake in Montana while he was setting in trout cages for a grant project he'd just been awarded. Given that Matthew left Earth so shortly after the great 1998 galactic alignment, this may have been something he could feel, since he was so sensitive to our planet. His wife Hillary and I believe that he *was* responding to that energy shift in the Galaxy in 1998.

Three weeks later, I received a fax from indigenous elders in the Yucatan, who did not know I'd lost my son. They wanted me to know that 12 Solar Ahau had left the Earth at the end of June and had flown out to the heliosphere to calibrate its energy fields, because Earth was shifting in response to a new galactic influence. I hope that Matthew is out there in that exquisite communication zone between our solar system and the Galaxy, the heliosphere. (The heliosphere is our solar system's membrane of life, the envelope around our solar system traveling through deep space.)

When Matthew died, my life went into a tailspin and I was unable to focus on my job acquiring books for Bear & Company. Soon thereafter, my husband Gerry, who was the president of Bear & Company, and I sold the company because neither of us could continue anymore. The hardest thing about letting my job go was that I was afraid it might prevent the crucial Maya codes from getting to the public in time. Academic publishers were not willing to publish the works of these highly speculative authors who were letting air into an academic field that was beginning to stagnate. As you can see, because of their commitment and greater resources since 2000, Inner Traditions/Bear & Company have already gone way beyond what Gerry and I could have accomplished, given our state of mind after losing Matthew.

In 2004, Bear published Carl Johan Calleman's *The Mayan Calendar* and sent it to me for my appraisal. I was stunned by the revelations in this book, and I immediately fired off a rave blurb. Then I sat down to contemplate the awesome implications of what Calleman had discovered about the Calendar.

What happened next is weird: I immediately sent a copy of *The Mayan Calendar* to my brother, Bob Hand. My forty-one-year-old son Tom, who worked with Bob, "borrowed," that is snitched, *The Mayan Calendar* from Bob. Tom was very curious about the Calendar because our Maya elder, Hunbatz Men, had given Tom the full Quiche Maya warrior codes during the "1989 Maya Initiatic Journey" in the Yucatan, a gathering for indigenous teachers. Tom and I were together on that 1989 journey, and I could see that the secret male Quiche initiations that he had undergone had made him into a spiritual warrior; he was very proud of this accomplishment, but he was not allowed to share the details, even with me, of what he'd experienced. When Venus crossed the Sun on June 6, 2004—a key event in the Mayan Calendar—Tom hanged himself from a tree. Before his body was found, somebody stole Tom's backpack. We are sure that *The Mayan Calendar* was in his pack, since Bob never did find his copy.

It cannot be meaningless that just before Tom took his life, he was reading Carl Johan Calleman's seminal book, which emphasizes the importance of spiritual access to greater consciousness during the Venus Passage. I have felt Tom's presence so strongly while writing *The Mayan Code* that I offer his story because I feel he has been writing it with me. Indeed, his soul may have chosen to go to spirit during the Venus passage so that he could exert a powerful influence from the other side. On a personal note, I'm Tom's Mom, and I knew him well: Tom chose to go to spirit because he'd had enough of earthly pain after losing his brother Matthew in 1998 and then his father, John Frazier, in 2001. Given all of this, however, I will always wonder what Tom was thinking about regarding Calleman's insights in *The Mayan Calendar,* and I honor his right to end his own life.

What's the point of my sharing this with you? You wouldn't be reading this book unless you are a person who is willing to consider definitions of reality wider than those espoused by the paradigms of scientific materialism, dogmatic history, and fundamentalist religion. What I have experienced in my life has changed me into a deeper person. As a teacher, I know how much each one of you has been challenged in your own way since 1998. I believe each of you has suffered as much pain as I have. Each one of us is going to die, just as Tom and Matthew have already, but that doesn't change the fact that you and I are alive during an amazing moment in time, the completion of the Mayan Calendar during 2011. We are all living in a time that requires great courage and insight.

Since 1998, I have often felt as if I was about to go insane, but each time I've managed to find my way back to sanity. What follows is a clear appraisal of how things work in the Earth plane, into which many dimensional realms fold, with the most complex dimension being time. As a result of channeling *The Pleiadian Agenda* in 1995 and analyzing its findings in *Alchemy of Nine Dimensions* in 2004, I feel especially equipped to speculate on the spiritual aspects of time and cycles.

In these two previous works, I explored the Mayan Calendar as both the generator of time and the time-driven evolutionary force from the ninth dimension. These findings were bizarre and incomprehensible to me until June 2005, yet always the ideas had a life of their own; they originated in the Pleiades. After all, the Pleiadians say that the ninth dimension *is* the Mayan Calendar! Well, I never imagined that another writer, Carl Johan Calleman, would come forth with a similar yet greatly expanded version of this idea of *sacred time*.

It is professionally important for me to state my own view of the correct date for the end of the Mayan Calendar. Many Calendar researchers are battling over whether 2011 or 2012 is the exactly correct end-date, and Mel Gibson has even jumped into the ring with *Apocalypto*. As you will see in this book, I believe Calleman has discovered the *purpose* of the Calendar—tracking evolutionary time acceleration through 2011. Yet I also believe that the equinoxes and solstices during 2012 (plus certain astrological factors explained in appendix B) will be a huge influence on humankind's ability to achieve enlightenment, so I have added that factor in detail.

Calleman's discovery of the vigesimal time acceleration is a new and huge accomplishment, which actually caused me to write *The Mayan Code*. (The vigesimal numerical system incorporates the concept of zero and is based on multiples of twenty.) Meanwhile, my own knowledge of astrology and historical cycles enables me to see the need for an analysis of the planetary influences during 2012, based on the assumption that Calleman's evolutionary theory is correct.

Thus, from my point of view, time and evolutionary acceleration will be completed on October 28, 2011, yet the Maya actually had the foresight to see that the Long Count will be completed on December 21, 2012, under the influence of critical astrological cycles. Astrology is little more than weather forecasting in the third dimension, yet we all like to know when a hurricane is coming. Time acceleration driving evolution is a force that functions in all nine dimensions, as I have explored in *The Pleiadian Agenda*. I believe that time in the ninth dimension is the actual driver of evolution in the Milky Way Galaxy, and I also think this concept is very close to what the Classic Maya thought. This idea of time may be the only theory that can explain why the Maya sacred year was 360 days long—as it was with many other ancient sacred cultures—but the agricultural year in the solar calendar was 365 days long, called the *Haab*. Discussing such ideas is only possible by acknowledging the considerable contributions of the great minds that have come before, which includes all the archaeologists who devotedly documented the glyphs of the Calendar and deciphered their meaning.

According to Calleman, June 2, 2005, was the midpoint of the period he calls the "Galactic Underworld," which began on January 5, 1999, and will continue until October 28, 2011. At the midpoint of any one of the Nine Underworlds, the evolutionary developments during that cycle become visible. The analogous midpoint of the great historical cycle from 3,115 BC through October 28, 2011 AD, was 550 BC, exactly when great teachers—such as Pythagoras, Zoroaster, Plato, Isaiah II, Lao-tse, Confucius, Mahavira, and the Buddha—appeared on the planet. The Maya most likely composed their Calendar around 550 BC, which means it is a script for enlightenment. The teachers of the Galactic Underworld have now—as of June 2005—appeared, so read on!

1

THE MAYAN CALENDAR

THE DISCOVERY

The story of the discovery and interpretation of the Mayan Calendar is one of dedicated and spirited scholarship by a cadre of adventurous people. These few intrepid researchers accomplished what was nearly impossible: they deciphered the Calendar's 5,125-year Long Count, and in so doing may have revealed the Creator's design for human and planetary evolution. Their legacy is probably the last opportunity for any of us to correctly interpret a highly advanced, ancient, sacred culture. During the 1930s, after enough dates had been deciphered and correlated, scholars could see that the Classic Maya (who lived from around AD 200 through 900) were obsessed with computing time and determining its meaning.

Working to decode the meaning of time, the Maya invented what is possibly the most sophisticated mathematical computational system in the history of human culture. Actually, the origins of this dating system considerably predate the Maya Classic period, and as you will see in a moment, there are many inscribed dates at Maya Classic sites that go back thousands, millions, even billions of years. These great dates are computed by means of a simple but brilliant system of bars, dots, and a symbol for zero: a totally accurate place-value notation system. The Olmec are the mother culture of Maya civilization, and elements of the Tzolkin—the 260-day count, which is still in use today—have been found at Olmec sites dating back around three thousand years, a further indication of the antiquity of the Maya dating system.

Most archaeologists believe the Olmec culture of central Mexico originated prior to 2000 BC. Be aware, however, that archaeologists of the last two hundred years are notorious for having underestimated the real ages of the origins of cultures. The archaeological data show that the Olmec were using the early vestiges of the Calendar; therefore, I begin this book by suggesting that the first arrival or congealment of the Olmec culture must have been around 3113 BC, the beginning of the Long Count. Since the Long Count describes the origins, development, and demise of Mesoamerican civilization, and because the Maya are the descendants of the Olmec, the seeds of the Maya must have already been sown by the very beginning of the Long Count.

In Maya mythology, the domestication of maize is associated with the origins of the people, and maize was discovered in the region seven thousand years ago.[1] The Calendar is called the Mayan Calendar (not Olmec) because the Classic Maya perfected all the aspects of it. They calculated how time influences history, and they left a clear and complex record of their discoveries. When they were developed enough to mythologize their origins, they linked the development of maize with their own origins in time in their creation account, the Popol Vuh. Today, corn remains a central focus of Maya culture and ceremony. The Long Count begins around 3113 BC, which happens to be the moment when complex temple-city civilizations suddenly arose in ancient Egypt, Sumer, and China. Since the Maya were also a culture with pyramids, hieroglyphs, mythology, and astronomy, and since their Calendar accurately describes the historical cycles of the development of civilization, why not assume their actual origins go back to the beginning of their Calendar, especially since maize was domesticated even earlier? This is an important point because all thirteen baktuns must be considered, especially the originating baktun, when complex civilizations began, which was 3113 to 2718 BC. (A baktun is a period lasting 394 years; baktuns are the main divisions of the 5,125 years of history that the Long Count describes.)

I will not spend much time on the pros and cons of scholarly disputes about Maya origins, since our interest is primarily in their Calendar. From my perspective as an indigenous person, I often differ with the opinions of archaeologists and anthropologists. I find that they are frequently wrong regarding the origins of cultures and their levels of complexity, especially concerning the native cultures of the Americas. In the case of the Maya, luckily time has already sorted out many of the early arguments about the Calendar, which makes it possible for me to add my own reflections. Calendar researchers are in remarkable concordance about the beginning- and end-dates—3113 BC to AD 2012, or 3115 BC to AD 2011 in the case of Carl Johan Calleman.

What few agree upon is what the Calendar actually *means,* such as the idea that it describes historical cultural evolution for 5,125 years, or evolution by time in general for 16.4 billion years, as Calleman posits. Academics rarely speculate about the meaning of the Calendar except to say that the Maya were obsessed with time. The writers I discuss at length in this book are not academics, yet they *are* scholars, and their interest is in uncovering the meaning of the Mayan Calendar and figuring out why the Maya were so obsessed with time.

In any event, we are all free to consider radical ideas about what the Maya might have been thinking, because modern cosmology has advanced enough to make this possible. We moved from a geocentric (Earth-oriented) to a heliocentric (Sun-oriented) perspective only four hundred years ago, right around the time that the last baktun of the Long Count began. Now, we are rapidly moving into a *galactocentric* (galaxy-oriented) perspective as this last baktun closes. It is only recently that most people have become aware that our solar system orbits around the center of the Milky Way Galaxy, and that our Galaxy is one among billions of other galaxies in the universe. Most researchers believe the Mayan Calendar tracks the development of culture by time, which is why people are so curious about it. We understand very little about the ancient Egyptian calendars because the limited Western materialist mindset controlled the first interpretations of this great sacred culture. More true interpretations of Maya culture are available because the discovery of the Maya goes back only 150 years, and most scholarship in this area has occurred during the last 100 years.

The public became aware of the findings of the early Calendar researchers in the 1950s, when scholars such as Sylvanus Morley and Eric Thompson published books about the numerical and mathematical systems of the Maya.[2] People were intrigued that the Maya used the vigesimal system (based on zero and multiples of twenty) for counting by dots and bars; few people could comprehend the awesome complexity of these great dates in the Calendar. The facts are that Maya science was incredibly advanced, and the concept of zero was not even *available* in Europe until the Muslims introduced it to the West in Spain around AD 1000.

CONTINUITY IN MESOAMERICAN CULTURE

The public was especially intrigued that millions of Maya descendants still live in villages near romantic pyramids in the jungle. Like ghostly guardians, the living Maya linger near cities that their ancestors had abandoned a thousand years ago. I will never forget my first ceremonies with the Lacondon Maya near Palenque in Chiapas, Mexico, in 1989. Like Essenes or Gnostics, the men wore pure white tunics of soft handwoven fabric that fell just below their knees, and their long, straight black hair fell down below their waists. Their faces were absorbed in another time when they brought forth pots they'd just recently made in the traditional way for the ceremony in the forest.

Isolated Maya clans in Chiapas in Mexico and in the highlands in Guatemala still keep the original calendars by counting the days. As well, they have preserved many of the most ancient rituals and healing skills. Luckily, since the 1960s and thereafter, sensitive scholars have gone to live among the Maya and document their ancient knowledge, a process so badly executed in the past by the conquistadors and Spanish clerics. Scholars who have actually lived with the Maya and learned their languages have added rich insights about the meaning of their original myths and dating systems. These insights have arrived just in time, for they put a check on the usual imposition of the Western cultural mindset. It would not be possible for more speculative researchers such as myself to consider and analyze the Mayan Calendar without the support of the findings of these seminal scholars at work in the field.

The Olmec, Maya, Toltec, Aztec, and living Maya continuity of art and mythology over great spans of time is impressive. The best word for this entire indigenous culture is *Mesoamerican,* since it honors the participation of living people in the ancient traditions. By the 1970s, a convergence of Maya studies was occurring because so many astonishing facts had been uncovered. Some of the most brilliant thinkers of our time—such as Terence McKenna and José Argüelles, who are best described as "new-paradigm researchers"—turned their attention to the Mayan Calendar. They have concluded that the Mayan Calendar is important for the Americas and for the whole world, and I agree. Some people, including myself, think that these Mesoamerican advances are the missing link that explains why the West has lost its soul during the last four hundred years. The Maya are the spiritual ancestors of people living in the Americas (as well as the Native Americans of the United States, the First Nation people of Canada, and the various ancient people of Central and South America), yet the meaning and viability of their traditions was denied and nearly eradicated.

Fifty years ago, growing awareness of the magnificence of Maya culture was opening great and wide perspectives in the history of ideas. For example, when Maya glyphs were finally deciphered, dates were found, such as at Coba in Southern Chiapas, that go back millions, billions, even trillions of years. In 1927, after these Maya dates were correlated with modern calendars—the GMT correlation (Goodman-Martinez Hernandez-Thompson)—most researchers agreed that the Long Count begins around 3113 BC.[3]

For many, it was amazing that the Maya had targeted the same date that archaeologists use for the simultaneous rise of civilization, which supposedly just emerged out of thin air with complex writing systems and elaborate temple cities popping up. New-paradigm writers have been commenting wryly that there are too many "suddenlys" in the historical dogma, and that something more is going on. I quite agree. If you think about it, the simultaneous global emergence of civilization in 3113 BC feels organic, as if humanity received an evolutionary signal 5,125 years ago. Well, as you will see, it did!

TIME CYCLES IN THE CALENDAR

What fascinates many is that the Long Count divides into cycles instead of just rolling along endlessly in linear time. As I get into this, Mayan cycles may not fascinate you at all, they may even seem to be as impenetrable as Greek. To help you absorb this part more easily, Chris has made some cartoon-like drawings, based on my sketches, that depict the basics of Mayan calendrics. In my case, I have to understand the Calendar mathematically to write about it, but you can just imprint it visually to read about it. Actually, that is how the Maya did it themselves, which is why it is likely that many of the people knew a lot about the symbols and numbers (of the Calendar) as they still do today.

The Long Count of 5,125 years is divided into thirteen cycles of about 394 years called baktuns; each baktun is divided into twenty cycles called katuns, which are composed of twenty 360-day-long tuns. Since the basic unit, the tun, is 360 days, time inches backward about five days during nearly a solar year (a solar year consists of 365 days), which yanks one's mind right out of linear time.

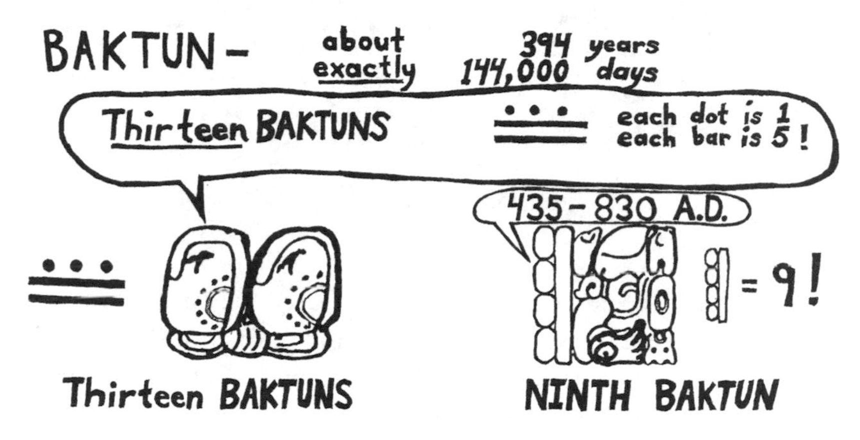

Fig. 1.1. Baktuns.

At major Classic sites such as Copan in Honduras, Palenque in Mexico, and Tikal and Quirigua in Guatemala—during Baktun Nine (AD 435 through 830 according to the Long Count) katun dates were often inscribed on steles, along with artistic and mythological depictions of historical and ceremonial events. Baktun Nine is only one section of time in the middle of the Long Count of thirteen baktuns, yet this was the period during which the Maya deepened their conceptions of what really goes on with time. We know that during Baktun Nine, ceremonies and divination were carried out constantly during important katuns and tuns. You could say *the Maya made time divine during Baktun Nine.*

The Classic Maya must have been studying the actual qualities of phases of time, since they erected and dated steles and paintings depicting content and mythology. They also depicted their mythology on thousands of Calendar-dated pots. In other words, they were telling their story and divining time. They seemed to have been attempting to discover the unique qualities of each katun (since 19.7 years was about one generation), which would then apply to the katuns of any future baktun.

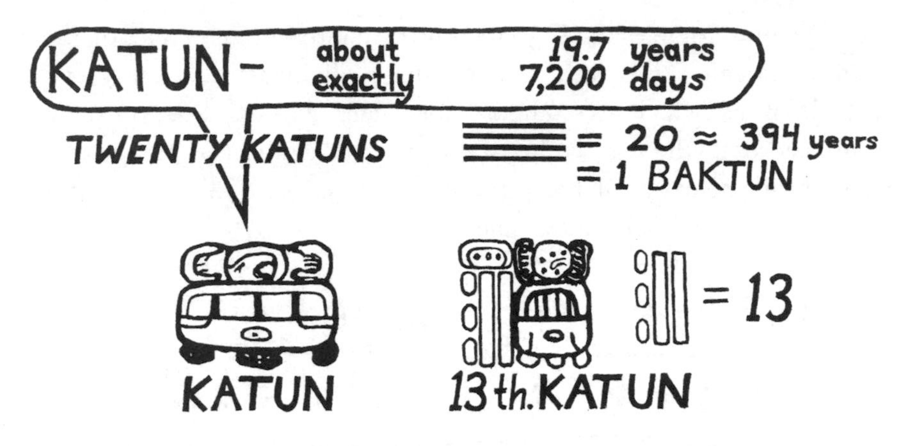

Fig. 1.2. Katuns.

The Tzolkin—the count of 260 days—is based on thirteen numbers and twenty day signs or symbols. In the Tzolkin, each of the day signs has unique qualities, which are altered by whatever number they happen to have. That is, 1 Kan is different from 8 Kan.

Fig. 1.3. Tzolkin.

The divisions that involve *qualities* are those of the twenty day signs (which are also each tempered by their numbers), which the Maya first discovered by experiencing the different days over a period of many years. Once they understood the twenty days and thirteen numbers—the Tzolkin of 260 days—then they explored the sacred year—the tun of 360 days—to understand qualities in larger and larger increments

Fig. 1.4. *Tuns.*

vigesimally, that is, time acceleration by 20 × 20. As you will see, the acceleration of time through the expression of multiplying basic units by twenty is a key concept in the Calendar, and it started with multiplying one tun by twenty to get one katun. Then a katun is multiplied by twenty to get a baktun, a baktun is multiplied by twenty to get a piktun, and so on. Therefore *the tun is the basic unit of time acceleration in the Calendar.*

The 360-day tun count is in resonance with the 260-day Tzolkin because the primary numbers that the Tzolkin is based on—thirteen and twenty—are both resonant with the longer thirteen baktuns (consisting of twenty katuns) which make up the Long Count. And just as each baktun has a quality of its own, which is numerical, so do the thirteen numerical days of the Tzolkin. I know this is confusing, hence it will be explained in more detail as we go along. What matters as we begin is to feel a sense of *resonance in time cycles*. To put it another way, a tun resonates or vibrates with a katun, a baktun, and so on. All this eventually leads right to the theory of vigesimal time acceleration.

The Classic Maya ended up with a very organic and mysterious view of time that still intrigues millions of people today. They seemed to have been trying to figure out how time grows and expands, which amounts to an attempt to capture the nature of evolution by time acceleration. As you will see when we investigate Carl Johan Calleman's theories, it is highly probable that they were investigating the acceleration potential of time. Think of how the minutes and hours drag when you wait to see your lover again, and think of how time flies once you are together. I think of Maya time acceleration as the essence of divine love because it fleshes out our personal relationship with the Creator; it describes the birthing potential of the universe.

VISITING MAYA TEMPLES

When I go to Classic Maya sites, I feel as though I'm in a living library inscribed in stone. The steles and inscriptions speak to me because they are sculptural, literary, and mathematical. If I can't figure out what I'm sensing, then while I am contemplating these messages, animals and insects reveal the codes to me by their behavior, which is why I always watch for signals from nature when I visit temple zones. The steles and inscriptions are guarded by palpable beings in other dimensions that joyfully open their secrets to me, or sometimes rudely push me away. Occasionally, I use incense or a rattle to contact the spirits guarding the stones. Once when I was rattling with my Turtle Nation rattle in the Temple of the Foliated Cross at Palenque, two arrogant archaeologists tried to interrupt me. They almost fell off the edge of the temple when invisible temple guardians pushed them!

I've sat for hours meditating with the steles, which sometimes finally come alive to reveal their records. In 1988 at Tikal in Guatemala, in a great court containing many steles, I could feel data flowing into my brain as if my cranial hard drive was being inscribed. I was so amazed by this that I asked for a sign that day that might help me believe what I'd just experienced. Later, while I was walking on a very remote path, a black jaguar appeared close by the path, stared into my eyes, and then slowly slunk away. This was especially bizarre because jaguars in the region are golden with black spots yet this black one seemed to be totally physical. I knew I'd have to trust what the stones said to me, which resulted in this book coming forth. It fascinates me that Carl Johan Calleman uses the main pyramid at Tikal as the model for his tun-based Calendar.

Like players in great dramas, today the steles stand mute year after year drawing our eyes, activating our bellies, and awakening our minds.

The more scholars investigated the dates and symbols of the steles, the more it seemed as if the Classic Maya were depicting cultural cycles in phases of time to indicate the meanings they'd discovered in these dates. Meanwhile, at the very least, the steles are fantastic works of art that communicate meaning just as successfully as a great painting by

Vermeer or a magnificent sculpture by Michelangelo; yet, of course, it's all a matter of what each person sees.

NEW-PARADIGM INTERPRETATIONS OF ANCIENT CULTURES

During the 1980s, a new group of scholars began speculating about the meaning of the Long Count and wondering about the scientific and spiritual accomplishments of the Maya. During this same time, the grip of the neo-Darwinian paradigm, which posits that humanity is constantly evolving to an ever-more-advanced level, was breaking down. The theory, as well as the dogma, that ancient people were less developed—while modern humans are becoming more and more advanced—was crumbling as the brilliance of ancient cultures was being revealed. It also became clear that archaeologists had, for the past two hundred years, misinterpreted ancient cultures. After looking at modern cultures with their eyes open, many people were having doubts about the supposed ever-progressive advancement of humanity.

People who were resisting neo-Darwinian projections longed for new interpretations of ancient cultures. New-paradigm writers—such as Graham Hancock, John Michell, and Peter Tompkins—responded to this curious public, and they found it easy to show that there were many ancient civilizations much more advanced than the current ones.[4] Meanwhile, academic archaeologists vigorously ignored and debunked this new paradigm and its followers in popular culture. In Maya research, the new paradigm arrived recently enough to allow researchers to discover that the Maya had developed a very advanced mathematical and astronomical system.

The new-paradigm concept of exploring advanced ancient wisdom flourished as the public became disenchanted with where barbaric modern civilizations, such as the United States, were headed; many people began to feel we needed ancient wisdom for the very survival of modern civilization. And, of course, we do! A rich literature, which sought to imagine the true potential of humanity based on ancient civilizations' greatest accomplishments, emerged fully matured in the 1990s. That, I believe, is what makes the Mayan Calendar urgently important now.

THE LEGACY OF THE MAYA

Now that we have entered the twenty-first century, the Mayan Calendar has seized the public imagination, partly because it is ending soon, and partly because it just fascinates people. Understanding the Calendar is very daunting, but as you will see in this book, we are in the middle of a quantum leap in decoding the Calendar. The Maya saved their Calendar in stone, whereas we have only vestiges of the calendars of other ancient cultures. You will see that the Maya carefully nurtured a legacy that may take us to a new level of understanding about the other brilliant legacies from the ancient Egyptians, Minoans, Sumerians, Chinese, and Vedic Indians. Although the Maya did employ a 365-day solar calendar for everyday use, which they called the *Haab,* the Mayan Calendar itself is all based on 360-day increments or cogs, called tuns.

If you look closely into the above-referenced ancient cultures, some of which suddenly emerged 5,125 years ago, they all have vestiges of a 360-day sacred calendar like the one based on the Long Count—especially the dynastic Egyptians and the Vedic culture of India—although many of them also have the agriculturally based 365-day solar calendar as well.[5] Nobody had been able to figure out why 360-day calendars were used in the distant past until Carl Johan Calleman began to study the matter. What concerns us here is that this calendar, based on 360 days, was saved in Mesoamerica while it was lost elsewhere. In other words, *the 360-day calendar reflects what was once a universal understanding of time.* As you will see, I think this says some very important things about the last ten thousand years of human experience. It is certainly noteworthy that the newly arrived Spaniards did everything possible to destroy this data; clearly, indigenous knowledge threatened them for some reason.

One of the purposes of this book is to figure out the reason for this division of time by 360, instead of 365, which was so important to the Maya two thousand years ago. For example, if Earth's journey around the Sun has lengthened by five days in relatively recent times, then what if that has changed the harmonics of Earth in the solar system and in the Galaxy? Will Earth have a tendency to return to a 360-day resonance in the future? If it did, how could that happen, and what would be the result? And could such a harmonic shift be what the end-date of the Mayan Calendar is all about?

THE CONTRIBUTIONS OF JOHN MAJOR JENKINS

In the late 1980s, once the information about the Mayan Calendar was sufficiently developed and published, nonacademic scholars began to offer highly speculative, exceedingly fascinating, and very creative interpretations of the Calendar. As mentioned in the introduction, I was involved in publishing many of these writers, and much of this book is devoted to their work as well as my own speculations about the Calendar.

I begin with John Major Jenkins, a brilliant new-paradigm scholar, because John made a pre–Classic Maya site, Izapa in Mexico, come alive for me in a way that fits with the intense feelings I've had myself at sites like Tikal in Guatemala, and Palenque and Teotihuacan in Mexico.[6] When I've been able to access the codes or central teachings of a site, suddenly I am in the middle of a mystery play, and everything comes alive like a movie: The film starts rolling as the frescoes take on their original hues, and the personages they depict open their eyes and move their lips. The steles download information, while animals and insects respond to the unfolding pageant and show me what it all means. Once, three Quetzal parrots swirled above my head when I was meditating in the Bat Temple of Tikal. Another time, a fer-de-lance serpent appeared right in front of me just when I fully comprehended the meaning of Manik, a day sign. These are the kind of great ecstasies I've experienced in sacred sites, and in *Maya Cosmogenesis 2012,* John Major Jenkins makes Izapa come alive on the same level. Because his scholarship is often better than most scholars who are published by university presses, I trust his insights, although I don't always agree with him.

My first enjoyment of *Maya Cosmogenesis 2012* came when I read the manuscript to consider it for publication. His book invited me to become an astronomer at Izapa, contemplating the zenith position of the Pole star or the Pleiades or the center of the Milky Way Galaxy in the Dark Rift. This was fun because I was a practicing astrologer at the time. I could imagine being a player in the ballgame at Izapa or giving birth on the cosmic throne. Amid his selection of key steles at Izapa, John leads the reader through the stages of the Izapa initiation systems of a few thousand years ago. I've never been to Izapa, yet his book took me right there and encouraged me to linger long enough to learn what I needed to know.

Jenkins has elucidated an astounding theory that Mesoamerica developed a two-part cosmology that works with the zenith positions of the Pleiades and the center of our Galaxy. These ideas have not been taken up much by Maya scholars, yet Jenkins mastered their work before moving beyond them into new territory. Scholars would be smart to pay close attention to his conclusions, which are based on their own painstaking work. For example, before *Maya Cosmogenesis 2012* was published in 1998, Linda Schele and David Friedel had attempted to bring Maya astronomy and ceremony alive in their *Maya Cosmos* in 1993.[7] Unfortunately, *Maya Cosmos* started academic Maya cosmologi-

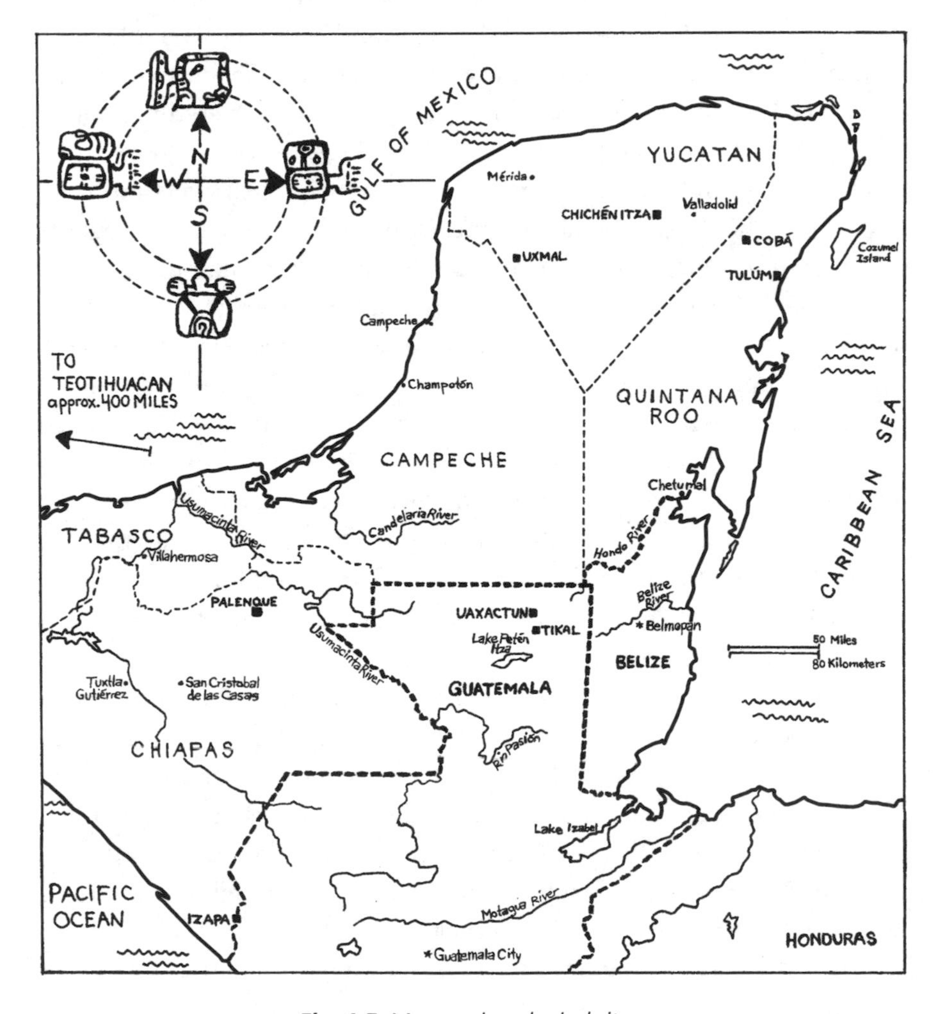

Fig. 1.5. *Maya archaeological sites.*

cal speculations down the wrong path by its focus on the Galactic *anti*-Center instead of the Galactic Center.[8]

I have already speculated that the world sacred cultures that suddenly appeared around 3115 BC used calendars based on systems similar to the Calendar of the Maya. In 1994, my Egyptian teacher, Abd'El Hakim Awyan of Giza, and I conducted ceremonies in the Tomb of Maya, in the Nineteenth Dynasty section of Saquarra near the Pyramid of Unas. Hakim created these ceremonies because he felt it was important for me to understand that the Maya and Egyptians were linked around 1200 BC, around the time the early stages of the Calendar showed up in Mesoamerica.

Hakim is the modern carrier of the Keys of Thoth, an Egyptian wisdom school that is five thousand years old. He says there are exactly 360 *Neters*—sacred keepers of the days—which is the same concept as the Maya tun-based year.[9] As well, this relates to the astrological Sabian symbols in that, as astrologers, we still use a 360-day year as our method for logging sacred time. I don't think the division by tuns, katuns, and baktuns was fully understood by the Maya until about two thousand years ago, and as I've said, they comprehended the full meaning only during the following thousand years. The oldest Long Count date (the earliest date found on a monument so far) is 37 BC at Izapa, but the elements of the system are much older.

We began this section with Jenkins' critical-leap understanding of how the Maya found the Long Count. To figure this out, Jenkins used archaeoastronomy—the study of how cycles in the sky reflect cycles on Earth. Archaeoastronomy uses the Hermetic principle "As above, so below" to determine how ancient cultures constructed their sacred sites using the positions of stars and constellations as a blueprint. Ancient sacred sites are oriented to positions in the sky during a specific period of time. We can date the sites by these orientations, and then we can speculate about the original people who built them by figuring out what they were focusing on; we can use their science to know them. Although archaeologists somewhat acknowledge the validity of this method, they often ignore the findings of archaeoastronomy, probably because they associate it with astrology. Yet, archaeoastronomy is simply ancient astronomy. As stated above, looking at space and time by means of archaeoastronomy leads to speculations about what people were really thinking about; in this line of investigation, John Major Jenkins excels.

In my opinion, his theory of how the Maya could have found the end-date of the Mayan Calendar is fundamental to understanding how they figured out when it begins in the first place.

THE FALL OF THE POLAR GOD

Jenkins finds that Izapan astronomers were first obsessed with observing the Pole Star.[10] Since the Pole Star is barely visible from Izapa in Southern Mexico, this obsession suggests that some if not all of their ancestors had come to Izapa many years ago from the far north. Ironically, this supports the conventional anthropological theory of the peopling of the Americas, although it is also likely that the diffusion theory, which posits that people came to the Americas by boat, is also accurate.

Still today, people in northern latitudes find their cosmic center in the Pole Star and the circumpolar stars. Since the Pole Star was the original cosmic center for the ancestors of the Izapans, they are a very ancient people. However, once they began to develop their culture in the tropics (between 23 degrees north and south latitudes) at Izapa, the changes in the locations of the circumpolar stars—caused by precession—were very unnerving; *their god had fallen in the sky.* Decoding their ceremonial sites, and taking the bible of the Maya—the Popol Vuh—into account, Jenkins next demonstrates how the Izapans tracked the zenith position (Zenith) of the Pleiades, as well as the Zenith of the Sun, until around 50 BC.[11] After that, they began to focus on the center of the Milky Way Galaxy.

What were they looking for? They were seeking a nonmoving, transcendent center in the sky to be their god, their sense of the divine in the universe. Being a people who mostly originated in the north thousands of years previously, their first god was the Pole Star because, from the perspective of the northern latitudes, everything seems to circle around the Pole.

At first, the Pleiades high in the night sky seemed not to move, which looks that way because the Pleiades (located near the Galactic anti-Center) are *opposite* the Galactic Center. When you look at the Galactic Center in Sagittarius, you are looking into the Galaxy from Earth, and when you look out to the Pleiades you are looking out of the Galaxy into the universe. Because the Pleiades appeared not to move, the Pleiades became their center, their god.

THE DARK RIFT OF THE MILKY WAY GALAXY

With more observation, the Izapan astronomers could see that not only were the Pleiades moving by precession, they were, in fact, approaching Zenith at Izapa. The place that never moved, however, was the Dark Rift of the Milky Way Galaxy, the place pointed to by the arrow of Sagittarius and the Scorpion's tail. Shamans would have traveled in their minds into this zone of the sky and must have felt time-reversals in the Black Hole, as well as new stars being born. The Galactic Center is the divine center for Earth! From Earth's perspective, it is the only stationary place in the sky, as well as being a place of awesome energy.

Earth's axis tilts by 23½ degrees as it orbits around the Sun, which causes the movement of the Sun on the horizon during the seasons. Meanwhile, where the Sun rises in the constellations is always changing by precession. The plane of the solar system, the ecliptic, slices the galactic plane at a 60-degree angle, which you can see from Earth when

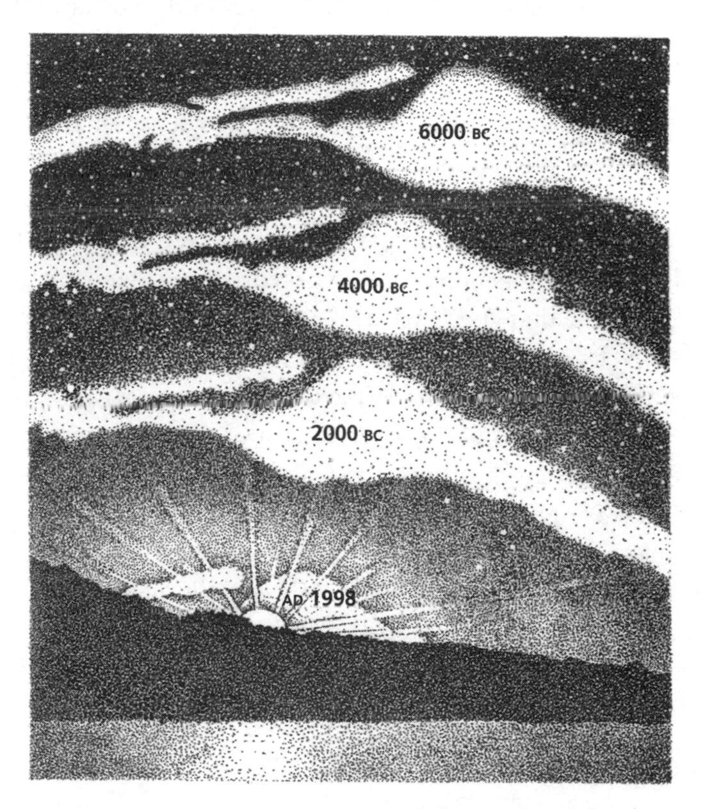

***Fig. 1.6.** The Milky Way's "fall" toward the winter solstice Sun from 6000 BC to AD 1998 as seen from Izapa, Mexico. (Illustration adapted from Jenkins'* Maya Cosmogenesis 2012.*)*

the edge of the Galaxy is visible as the planets and the Sun traverse the ecliptic. This is a dramatic tilted cross in the sky when viewed in the Tropics, where I live part of the year. Sixty-degree angles are the six angles of the famed Star of David, and I think this beautiful star is a galactic symbol.

Two thousand years ago, the Izapans noticed that the rising winter solstice Sun was moving closer to their sacred center, the Galactic Center in Sagittarius. Their god, the Sun, the source of all life on Earth, was moving toward their cosmic center! Next, according to Jenkins, by calculating precession, the Izapans determined that the rising winter solstice Sun was going to conjunct the Galactic Center approximately two thousand years in the future.[12] This was very visible to them, as you can see in the illustration adapted from *Maya Cosmogenesis 2012*. The precessional calculations of ancient stargazers were very accurate, and the Maya were no exception. Jenkins posits that, when it dawned on them that this conjunction would occur around two thousand years in the future, the Izapans used their own numerical system (Tzolkin) to construct the Long Count based on this end-date.[13] I also believe they had a sense of the date of their origins, and this sense of the beginning and end of history offered them astonishing revelations. But I am getting ahead of myself. Once the Maya began to construct the Long Count, they began dating stele with Long Count dates. Yet, these dates were in the middle of their Calendar. They worked out a calendar targeted at the 2012 end-date that goes back 5,125 years to the time of their origins within this cycle. Isn't that amazing! The Long Count encompasses their origins, development, and completion over a span of 5,125 years.

Since many other complex civilizations appeared simultaneously with the Maya, it is likely that both the Calendar and the end-date of the Long Count have global consequences. Why? What could this mean to you and me? Can you see why some researchers have almost lost their sanity trying to figure this out? Well, as you will see when wc consider what Carl Johan Calleman has to say, *this end-date may apply to 16.4 billion years of evolution on Earth that is culminating in 2011!*

THE MAYA OBSESSION WITH TIME

Being in the middle of these final stages of Mayan Calendar study since the 1970s, a field that is famous for driving researchers crazy, I have been just as obsessed as anyone with the meaning of the Calendar. All I've had to go on besides continual deep study has been my involvement as a medicine woman and the use of my intuition, which has often been the best guide of all. What I know comes from hours of meditation with the stones in Mesoamerica and during ceremonial work, and from the discoveries of others. Usually what I've intuited has been verified eventually, so I keep track of things I feel are right that everybody else dismisses. I'd like to offer you more of a sense of my own perspective, which ultimately is very intuitional, since it is the underpinning of this book. If you are seeking the Logic 101 of the Mayan Calendar, it's not here.

Regarding my own intuition, my primary guides are my Cherokee-Celtic grandfather Hand, who gave me a *time-coded legacy* (a description of history and evolution based on time cycles), and the subsequent verification of this legacy by Hunbatz Men of the Yucatan, who was also given a legacy from his family. Abd'El Hakim Awyan of Egypt was given his legacy as Keeper of the Keys of Thoth. I was taught as a very small child and into my adulthood that I must remember all the records I was given. I was expected to find a way to give the records back to the people during the end of the Calendar. I was taught to be obsessive about this legacy when I was very small, and during the bland 1950s in America, my parents did not like Grandfather's influence on me at all. Grandfather convinced me that if I deviated from this quest, I would lose touch with my source, the Pleiades star system. My Scottish grandmother agreed with him, and they were the primary influences in my life.

Along with the Maya, the Cherokee retain aspects of the Calendar, yet they did not incise the data in stone or keep the day count for the past twenty-five hundred years. The only real clue I got about the Galactic Center was that we Cherokee have a great crystal turtle that teaches the people about the Galactic *anti*-Center, since in the sky, Orion is Turtle's back, where the people of Turtle Nation (North America) came from. When I first began working with Hunbatz Men in the 1980s, I was greatly relieved to hear him constantly talking about the Galaxy and the Pleiades, since that topic of conversation ended for me when Grandfather Hand died in 1961. The Orion constellation is very close to the Pleiades and the Galactic anti-Center. Orion and the Pleiades are like Greyhound bus stations for shamanic travelers leaving the Galaxy, including me. Look up in the night sky and notice the opening in the edge of the Galaxy near Orion and the Pleiades.

As an adult, I was also taught to pray with twenty prayers, four to each direction plus a fifth in the center—the prayers for all the plants, animals, minerals, and beings in the universe. These Cherokee prayers are similar to the Maya counting system. The most important portion of Grandfather's legacy that I can contribute to this book is his teaching about a cataclysm 11,500 years ago that may have actually generated the need for the people to have calendars in the first place. I've already thoroughly explored this great cataclysm in my book *Catastrophobia,* yet because it may be the very reason the 360-day Calendar was invented, I will summarize it again later in this book. It has significant information about what the Mayan Calendar describes, which I couldn't see myself until I first understood Carl Johan Calleman's work in June of 2005.

HARMONIC CONVERGENCE: AUGUST 16–17, 1987

So there I was in the 1980s, working as a publisher and aware of the Galaxy as center. Then the Lakota teacher Tony Shearer informed me how important August 16–17, 1987, would be for indigenous people. He said this date was the end of the Nine Hells and the beginning of the Thirteen Heavens in the Aztec Calendar, which meant the end of the eradication of the wisdom of Mesoamerica.[14] I didn't follow the Aztec Calendar, but I sensed this was important. Soon thereafter, José Argüelles came to my office with *The Mayan Factor,* a book about August 16–17, 1987, which was an event he named Harmonic Convergence. Bear & Company quickly got it ready for publication by early 1987, while I studied it diligently.[15]

We entered into pre-Harmonic Convergence ceremonies when Gerry and I went to Palenque with Argüelles and a few people in February 1987. Just as Supernova 1987 was depositing neutrons deep in the Earth—a cosmic effect confirmed by scientists—José, with his great shamanic powers, made the Court of the Nine Bolontiku (the main ceremonial court by the Palace at Palenque) come alive.[16] *The Mayan Factor* discusses a "Galactic Synchronization Beam" that Argüelles says has been activating Earth during the whole 5,125-year Long Count.[17] It was especially important that people go to sacred sites to do ceremonies to connect with this beam on August 16 and 17, 1987.

During Harmonic Convergence, Gerry, our son Matthew, and I went to Teotihuacan—"where the gods come down to Earth"—the great

Toltec city north of Mexico City. The culmination for us was a sacred pipe ceremony with White Eagle Tree, my Cherokee medicine brother, in the Temple of the Quetzal Butterfly, and the ceremonies were beautiful. What happened with all the indigenous people that day forever altered my understanding of how much our planet is really changing. Before dawn on the morning of August 17, our small group of one hundred or so individuals who would participate in the ceremonies went to the temples. Over one hundred thousand Mechica (Mexican Indians) who were eagerly swarming into Teotihuacan for the ceremonies almost ran us into the ground; for days, they had been walking there from all over Mexico.

Harmonic Convergence gatherings like this one were a global phenomenon. And, during key times since 1987, Mexicans have continued to hold ceremonies of this type. The media satirized this event. For example, because my old friend Garry Trudeau thought what I was doing was funny, he used his *Doonesbury* cartoon to depict "Boopsie" flying over Teotihuacan during Harmonic Convergence. *The Wall Street Journal* panned the events, and on *The Tonight Show,* Johnny Carson snickered about Harmonic Convergence. The more the media laughed, the more Americans (who normally would never hear about indigenous ceremonies) became aware of this important date. Millions of people around the world spontaneously responded to this event. Since I was used to doing very secretive ceremonial work, this astonishing event tuned my consciousness for the first time to the real potency of the galactic influence on the wider public. Besides helping me begin to comprehend the galactic influence, the other thing that impressed me about *The Mayan Factor* was José Argüelles's historical analysis of the Long Count's thirteen baktuns, the 394-year-long developmental stages from 3113 BC until AD 2012. Argüelles discovered that the Long Count describes a long wave of history that tracks the rise and fall of civilizations all around the globe.[18] His analysis is thought provoking.

His conclusions led me to wonder why the Maya would be interested in tracking the history of cultures around the world. I contented myself with thinking about the various cycles. Certainly, the thirteen baktuns describe cycles of history during the last 5,125 years that are culminating in a rising wave of materialism that threatens the planet. This would suggest that as this long cycle ends, materialism must dissipate to make space for other ways to create human realities. These ideas led me to write my 2001 book, *Catastrophobia*, which I believe is an essential element in this discussion.[19] The mentality created on Earth by the cataclysm in 9500 BC greatly altered the minds and emotions of humanity, and the clearest accounts of this pain are found in the records of Mesoamerica and in Turtle Nation, the home of indigenous Native Americans and First Nation people in Canada.

Today, for a rapidly increasing number of people, Carl Johan Calleman's evolutionary theories of the Mayan Calendar are very compelling. In order to investigate that further, I will end this discussion about the sources of the Mayan Calendar, and we will shift to speculations about what the Calendar may actually mean. The next chapter focuses on Calleman's two books on the Mayan Calendar.

2

ORGANIC TIME

EVOLUTIONARY THEORY AND THE MAYAN CALENDAR

Carl Johan Calleman, who graciously wrote the foreword for this book, is a Swedish biologist who began studying the Mayan Calendar in 1979. By using the Mayan Calendar as a template for the study of biological evolution, he noticed what he calls a "rather simple and telling pattern."[1] He could see that the Second Underworld of the Calendar, the so-called Mammalian Underworld, contained a series of thirteen dates (the Thirteen Heavens) that describe the major transitions of biological evolution during 820 million years, the period when Earth's multicellular animals developed. Amazingly, the previous and first cycle of 16.4 billion years in the Calendar (which is exactly twenty times longer than the 820 million-year Mammalian cycle) is very close to when cosmologists theorize the universe was created.

Others had noticed that the great numbers recorded by the Classic Maya in their Calendar are in close concordance with the transition dates of cosmological and evolutionary science, but it took a biologist to see the incredible significance of this fact. For Carl Johan Calleman, figuring out the meaning of these matching databanks has become his life's work. He wondered how the Classic Maya could have known about the evolutionary cycles, since modern science only discovered them during the last two hundred years and moved into agreement on this data during the last fifty years.

Ruling Energy	National Underworld 13 *baktun* 5,125 years 13 Days/Nights of 394.3 years	Planetary Underworld 13 *katun* 256 years 13 Days/Nights of 19.7 years	Galactic Underworld 13 *tun* 12.8 years 13 Days/Nights of 360 days	Universal Underworld 13 *uinal* 260 days 13 Days/Nights of 20 days
Day 1 is Heaven 1 **Sowing** *Xiuhtecuhtli,* god of fire and time	Aug. 11, 3115–2721 BC	July 24, 1755–1775	Jan. 5, 1999–Dec. 31, 1999	Feb. 11, 2011–March 3
Night 1 is Heaven 2 **Inner Assimilation of New Wave** *Tlaltecuhtli,* god of earth	2721–2326	1775–1794	Dec. 31, 1999–Dec. 25, 2000	March 3–March 23
Day 2 is Heaven 3 **Germination** *Chalchiuhtlicue,* goddess of water	2326–1932	1794–1814	Dec. 25, 2000–Dec. 20, 2001	March 23–April 12
Night 2 is Heaven 4 **Resistance against New Wave** *Tonatiuh,* god of the sun and the warriors	1932–1538	1814–1834	Dec. 20, 2001–Dec. 15, 2002	April 12–May 2
Day 3 is Heaven 5 **Sprouting** *Tlacoteotl,* goddess of love and childbirth	1538–1144	1834–1854	Dec. 15, 2002–Dec. 10, 2003	May 2–May 22
Night 3 is Heaven 6 **Assimilation of New Wave** *Mictlantechutli,* god of death	1144–749	1854–1873	Dec. 10, 2003–Dec. 4, 2004	May 22–June 11
Day 4 is Heaven 7 **Proliferation** *Cinteotl,* god of maize and sustenance	749–355	1873–1893	Dec. 4, 2004–Nov. 29, 2005	June 11–July 1
Night 4 is Heaven 8 **Expansion of New Wave** *Tlaloc,* god of rain and war	355 BC–AD 40	1893–1913	Nov. 29, 2005–Nov. 19, 2007	July 1–July 21
Day 5 is Heaven 9 **Budding** *Quetzalcoatl,* god of light	AD 40–434	1913–1932	Nov. 24, 2006–Nov. 19, 2007	July 21–Aug. 10
Night 5 is Heaven 10 **Destruction** *Tezcatlipoca,* god of darkness	434–829	1932–1952	Nov. 19, 2007–Nov. 13, 2008	Aug. 10–Aug. 30
Day 6 is Heaven 11 **Flowering** *Yohualticitl,* goddess of birth	829–1223	1952–1972	Nov. 13, 2008–Nov. 8, 2009	Aug. 30–Sept. 19
Night 6 is Heaven 12 **Fine tuning of New Protoform** *Tlahuizcalpantecuhtli,* god before dawn	1223–1617	1972–1992	Nov. 8, 2009–Nov. 3, 2010	Sept. 19–Oct. 9
Day 7 is Heaven 13 **Fruition** *Ometeotl/Omecinatl,* Dual-Creator God	1617–**Oct. 28, 2011**	1992–**Oct. 28, 2011**	Nov. 3, 2010–**Oct. 28, 2011**	Oct. 9–**Oct. 28, 2011**

Fig. 2.1. *A chart of prophecy—the Calleman Matrix—illustrates the ruling periods of the thirteen deities in the National, Planetary, Galactic, and Universal Underworlds. (Illustration from Calleman,* The Mayan Calendar and the Transformation of Consciousness.*)*

Before we explore this wonderful question, let me point out that the Mayan Calendar, which actually does describe vast stretches of time that may match the evolutionary databank, is why the Maya are of such compelling interest right now. I will not be spending much time on Maya culture and archaeology in this book, since there are so many great books already written on these subjects. The long dates of the Calendar are of critical importance because they apparently describe the evolutionary processes of the universe, and also because the Calendar ends in only a few short years. What if this ending says something about the process of evolution itself? What if a phase of evolution or the completion of evolution itself is what the Calendar describes?

Also, we must have great respect for any knowledge that the indigenous people retained in the face of the destruction of their culture during the last four hundred years. Maya descendants in Guatemala and Mexico protected remnants of the Calendar by keeping the count of days—the Tzolkin—for at least 2,500 years. Contemporary Maya did not retain the concept of the Long Count (3113 BC–AD 2012), as far as we know, thus our knowledge of the Long Count comes from inscriptions and books created by the ancient Maya.

As you will readily see in this chapter, Calleman's Thirteen Heavens of the Mammalian Underworld actually do describe how the evolutionary force expresses itself in nature. Regarding all Nine Underworlds, Calleman's theory is a brilliant one that may explain why we have arrived where we are as a species. My consideration of Calleman's hypothesis of biological evolution and the Mayan Calendar is the central theme of this book, since I am convinced he has discovered the real meaning of the Calendar. If you are interested in biological evolution, the Calendar, or both, I highly recommend that you read his work to fully understand this discussion.

THE NINE UNDERWORLDS OF CREATION

In the previous chapter, I mostly touched on the sixth of the Nine Underworlds, the 5,125-year historical cycle that Calleman calls the National Underworld. In scholarly sources, this cycle is the same as the Long Count or the Initial Series, which is 3113 BC–AD 2012. The great dates, such as 16.4 billion years, are simply multiples by twenty of elements in the Long Count, which means the Calendar is a vigesimal system that uses units reckoned by twenty. As already noted, Calleman calls the Long Count the Great Year because in his system it corresponds to the 5,125-year-long cycle in all ways *except* that it begins in 3115 BC and ends in AD 2011.

From now on in this book, I will use Calleman's dating for all the cycles in the Mayan Calendar, rather than switching back and forth from his Great Year to the Long Count. As you read on, it will be obvious why I am using his dating system, but you need not be concerned about this, because the differences are so small, barely a year. As you will see, if Calleman's dates are correct, the differential in the dates in the past is inconsequential, but the difference of one year will be a big deal in 2011. Regarding why Calleman uses a slightly different end-date, he believes the Classic Maya of Palenque tweaked the date slightly around a thousand years after the Long Count was devised.[2] In this chapter, we investigate Calleman's Nine Underworlds, which encompass 16.4 billion years of time. I will look for evidence that the Classic Maya actually might have thought along these lines themselves, or at least that they stumbled upon the "skeleton of time" that drives evolution. It certainly looks like the Maya were actually aware of the time cycles that go back to the creation of the universe, an idea cosmologists discovered and comprehended only very recently. If you can grasp the scope and accuracy of the Mayan Calendar dates, you will see the Maya were way ahead of modern science until only a few years ago. Meanwhile, the biologist Calleman's evolutionary theory—*that a time-coded force drives evolution*—is an incredibly radical thought. Therefore, we must first place this discussion in the context of current evolutionary theory, which is a contentious issue within popular culture in the United States at this time.

DARWINIAN EVOLUTIONARY THEORY VERSUS INTELLIGENT DESIGN

Darwinian evolution theorizes that random and undirected forces have evolved the universe by a mechanism called natural selection. Intelligent design posits that a conscious force (some call it God) exists that designed and guides all matter, including the process of natural selection. Since all of nature exhibits geometrical laws and orderly complexity, this is a perfectly reasonable possibility. Think of the geometry of seashells, hurricanes, and sunflowers. According to Chaos Theory, even *disorder* is not random. Europeans have widely accepted Darwin's theories only within the past fifty years, yet they realize that the theory of intelligent design does not contradict the laws of natural selection. Conversely, in the United States battles are raging in local courts over the legality of schools teaching *both* evolutionary theories.

In the United States, an intense debate over evolutionary theory is fueling a rapidly growing Judeo-Christian fundamentalist movement. Still clinging to theological theories about creation, possibly half the population believes that God created the world only six thousand years ago. Regardless of public opinion, during the last fifty years most American public schools have taught Darwinian evolutionary theory, and by so doing have created a severe mental dichotomy and resulting societal tension. As an example of how this has manifested, many fundamentalists have removed their children from the public schools and are educating them at home.

Reactions against Darwinian evolutionary theory have been building steadily among American fundamentalists, who have very literal minds. People in general, including some fundamentalists, have adopted the scientifically credible evolutionary theory of intelligent design, which postulates a structured and orderly process of evolution over billions of years that must be guided by some form of intelligence. However, most fundamentalists believe that intelligent design is also "Creationism," which posits the world was created six thousand years ago based on the Judeo-Christian calendars.

Theories of the intelligent design of the universe over the great spans of time described by science are as valid as Darwinian evolutionary theory, possibly more so. Obviously the universe has evolved over billions of years, yet this does not mean Darwinian theory or all the elements of scientific dating are correct. As you will see in chapter 3, I differ very much with the conventional timeline covering the last forty thousand years, possibly the last hundred thousand years. Regardless, Darwin and modern science are mostly right about the long cycles of time and the fact that humans evolved from hominids that evolved from monkeys. However, Darwinian theory takes the concept of natural selection—survival of the fittest determines reproductive success—too far. There is more going on in the process of evolution than just the survival of the fittest and random selection. Most significantly, natural selection focuses too much on physical evolution and grossly downplays the role of consciousness. In the theory that espouses evolution by intelligent design, consciousness is the driving force of evolution, and the physical changes follow creative intentions.

Going back to Creationism for a moment, I suggest that fundamentalists should consider the new evolutionary theories recently found by Calleman within the Mayan Calendar. His arguments provide powerful evidence for divine guidance throughout very long cycles of time. His interpretation of the Calendar challenges us to see that *only an awesome consciousness could have created the Mayan Calendar as well as the universe*. I think what the fundamentalists are uncomfortable about is the idea of humans evolving from animals, yet why should this offend them if their newly arrived God created everything? Considering all the effort the ancient Maya went to, they might have realized we'd need their calendrical knowledge now. Maybe they thought that their descendants—us—would be able to co-create with evolution by working with time acceleration near the end of the Calendar once we discovered time acceleration and its timing in the Calendar? That is, we would become conscious of our ability to be active participants in the development of life itself. As you will see, the Mayan Calendar is the most advanced theory of intelligent design that exists, yet its vast mind does not require the Judeo-Christian God—Yahweh—who, by the timing of the Calendar, has been on the stage for only a small amount of time (during the 5,125-year-long National Underworld). Although the Calendar is based on the premise of a great Creator, this is not the same as the concept of gods in time and history, such as Yahweh.

Those who are tired of the silly public arguments over whether Creationism or Darwinian theory is correct might feel they don't want to think about what Calleman theorizes. Some might even dismiss these new and complicated possibilities by concluding that Calleman is a closet fundamentalist. Therefore, I will clarify what fundamentalism is and is not. This is important because some people think the Mayan Calendar is an end-of-the-world scenario in 2011 or 2012, which it is not. Waiting for the end of the world is what I call *catastrophobia*—fear of catastrophe. When I wrote about this syndrome in *Catastrophobia,* I showed how it has caused humanity to become a multitraumatized species during the last 11,500 years.[3] This primary phobia blocks the human mind, reduces intelligence, and has profoundly retarded the progress of human evolution. Since most fundamentalists are afflicted with catastrophobia, their fearful mindset is capable of eclipsing the joyful anticipation of millions of indigenous people who are looking forward to the Calendar's end. For example, I have been rather anxious that fundamentalist cults, such as the Jehovah's Witnesses, might try to corner the market on the Mayan Calendar end-date by setting up 2011 or 2012 as their next end-of-the-world date.

AMERICAN JUDEO-CHRISTIAN FUNDAMENTALISM

What exactly is fundamentalism? People who adopt the Judeo-Christian fundamentalist approach (hereinafter *fundamentalist*) to theology base all their interpretations of reality on the Bible, yet they are neither traditionally religious nor conservative. They are radicals who interpret the Bible any way they want, and then they claim their interpretations are literal truth. Sadly, most Americans are so poorly educated in mythology, theology, and ancient history that they get excited when fundamentalist preachers talk about biblical stories of humanity's origins.

The arrogant and simplistic approach of the fundamentalists has been driving me crazy for years because I have a master's degree in theology from Matthew Fox's Institute for Creation-Centered Theology, and I love the Bible. Like the Popol Vuh—the mythological story of Maya origins—the Bible is a complex mythological story of the emergence of the Judeo-Christian people in history. People long to know this story because they live in a culture based on scientific materialism, which is devoid of meaning and devalues the past. The resulting agnostic American culture is so boring that, during the last thirty years, it has been easy for the fundamentalist preachers to capture the market on the stories of origin. It is unfortunate, really, and I have always felt deep compassion for fundamentalists, who are often afflicted with a potent and serious mental addiction to God.

Second, and much worse, once the fundamentalists have interpreted the Bible their own way, they apply these interpretations to contemporary society. They craft end-of-the-world scenarios, which stir up apocalyptical frenzy and insecurity in the public. The apocalyptic scenarios are used to justify wars against the "Infidel" because we are in the end-times. Fundamentalism strikes deep nerves in Judeo-Christian believers, since Jesus was born in an apocalyptic culture in which the Jews awaited the Messiah and the end of the Roman oppression. The end of the world never happened, Rome triumphed, and many Christians and Jews are still waiting for the Messiah. This tendency to live outside of current reality and be frothed up by all kinds of mythological obsessions makes it difficult for anybody to retain their sanity, much less their intelligence. Judeo-Christian fundamentalism should be identified as a new religion—"Apocalyptic Christianity"—which is the natural result of fear-filled reactions to the rapid levels of change going on since the mid-1700s. More and more people seem to be losing their critical faculties as they struggle to resist learning about the amazing scientific story of the universe that has emerged during the last two hundred years; *fundamentalism feeds on fear of change.*

INTEGRATING THE NEW SCIENTIFIC PARADIGM

In Calleman's system, this tense period when the scientific data bank emerged is the Seventh Underworld, the Planetary Underworld (AD 1755 to 2011), when a great speedup in human consciousness is taking place. According to the laws of the Calendar as interpreted by Calleman, evolutionary changes during this period are twenty times faster than those occurring during the National Underworld—3115 BC to AD 2011. During the 256 years of the Planetary Underworld, humanity has been developing planetary consciousness—globalization—triggered by the rise of industry. Often referred to as the European Enlightenment, this was/is the period when democracy came forth, which has been developing all the way through the cycles of the Planetary Underworld. During this head-spinning speedup, old forms of rigid social control that developed during the 5,125-year-long National Underworld have been breaking down. However, during the Planetary Underworld, the agenda was not really democracy, which was merely an idea that was used to distract people from the bigger economic program. The truth is, the people who were divided into castes during the National Underworld became the cogs in the industrial machine during the Planetary Underworld. Democracy did introduce the idea that people should have the right to individual freedom no matter what their economic status. Ironically for some, this aspect of democracy was and still is a threat, which has fueled the rise of fundamentalism. Freedom forces people to think straight, which is impossible for those who can't handle freedom and are afflicted with mental rigidity.

Another speedup by twenty times began in 1999. The Eighth Underworld—the Galactic Underworld—began January 5, 1999, and is causing an even faster speedup in evolution—twenty times faster than the Planetary Underworld, which is twenty times faster than the National Underworld. Take a moment to think about how time has accelerated during the Planetary Underworld with the rise of industry during a mere 256 years. Yet that is twenty times *slower* than the Galactic Underworld of 12.8 years (see below), in which the rate of change is measured in billions of cycles per second—gigahertz—and during which so much more can happen in so much less time. As I write, we are just past the midpoint of the Galactic Underworld, when exponentially fast change is intensifying our unprocessed fears and emotions. Remember, all three of these Underworlds are occurring simultaneously, as are all nine, since they all end in 2011. That fact is the great mystery discovered by Carl Johan Calleman. During the Planetary Underworld, a span of 256 years, people have gone from being caught in a hierarchical caste system to being cogs in the machine; now people are little more than walking credit cards constantly threatened with identity theft! And nobody wants to hear anybody say they are descended from monkeys, since many feel like monkeys when they try to handle modern technology.

In chapter 6, I present detailed information about the National, Planetary, and Galactic Underworlds. At this point, you may be resonating with the *idea* of these cycles, especially if you know something about recent history. To fully comprehend what Calleman is saying, we will first look at the longer and slower Underworlds in the Calendar. (If anybody is wondering where the term Underworld came from, it is simply a Maya term that Calleman adopted to connote the lower world that drives the evolutionary forces in our world.) We need a feeling for the long, slow times in our evolution before we deal with the faster ones, which are so confusing to most people today. Also, during rapid change, contemplation of the past can be very soothing, which is why so many people are obsessed with the past right now. We will consider 16.4 billion years of evolution during the first five Underworlds. Unlike neo-Darwinian evolutionary theory, the credo of scientific materialism, the Calendar does not see evolution as a series of random events in a godless universe.

The Mayan Calendar encourages us to see that everything that exists is an expression of a profound and ordered intelligence; everywhere there are traces of a great mind that does everything with a purpose. Darwin was right about the great spans of time, but he was only partially right about how it all works, especially as interpreted by his fanatical followers, the neo-Darwinians. Let's look at some long spans of time through the eyes of the Classic Maya of AD 200 to 900.

THE COBA STELE AND THE NINE UNDERWORLDS

At Coba, a Maya site in the Yucatan, there is a tall, thin stele describing trillions of years of time. As can be seen in figure 2.2, only part of the tuns—multiples by twenty—are named, such as baktun, piktun, kalabtun, and so forth. Above the last-named cycle, the hablatuns, there are fourteen unnamed powers of twenty.

The named cycles describe evolutionary cycles over 16.4 billion years. Cosmologists say the universe was created about 15 billion years ago, yet they might be more accurate to say 16.4 billion years ago. The other dates on this stele are either the same as the dates that scientific theory gives for the major evolutionary transitions, or they are very close. Yet it is only in the past 150 years that geologists and biologists have begun to realize that human beginnings have a long-distant past. Darwin published his *Origin of Species* in 1859, but the real advances in the study of hominid culture began during the mid-twentieth century at Olduvai Gorge in Tanzania with the work of Louis and Mary Leakey. The Coba stele dates were deciphered by the 1950s, so these matching databanks were discovered during parallel time frames, which I think is fascinating. These facts are indisputable: The Maya of Coba carved this stele around 1,300 years ago, and the dates on it exactly match modern scientific theory, which came together only about fifty years ago.

On stele 1 at Coba, the Long Count (5,125-year cycle) was put in the middle of named cycles that are each multiplied by twenty, which made it possible for the Maya to compute great spans of time that predate the period beginning 16.4 billion years ago. What on Earth could these great dates possibly have meant to the Maya? As Calleman puts it, here creation is regarded as a composite of creations, or cycles of evolution, and they are all built up on top of each other, which you can see in figure 2.3.[4] Each of these nine cycles or Underworlds is divided into thirteen Heavens, and the Long Count of thirteen baktuns (which Calleman calls the Great Year) is only one of these great cycles.

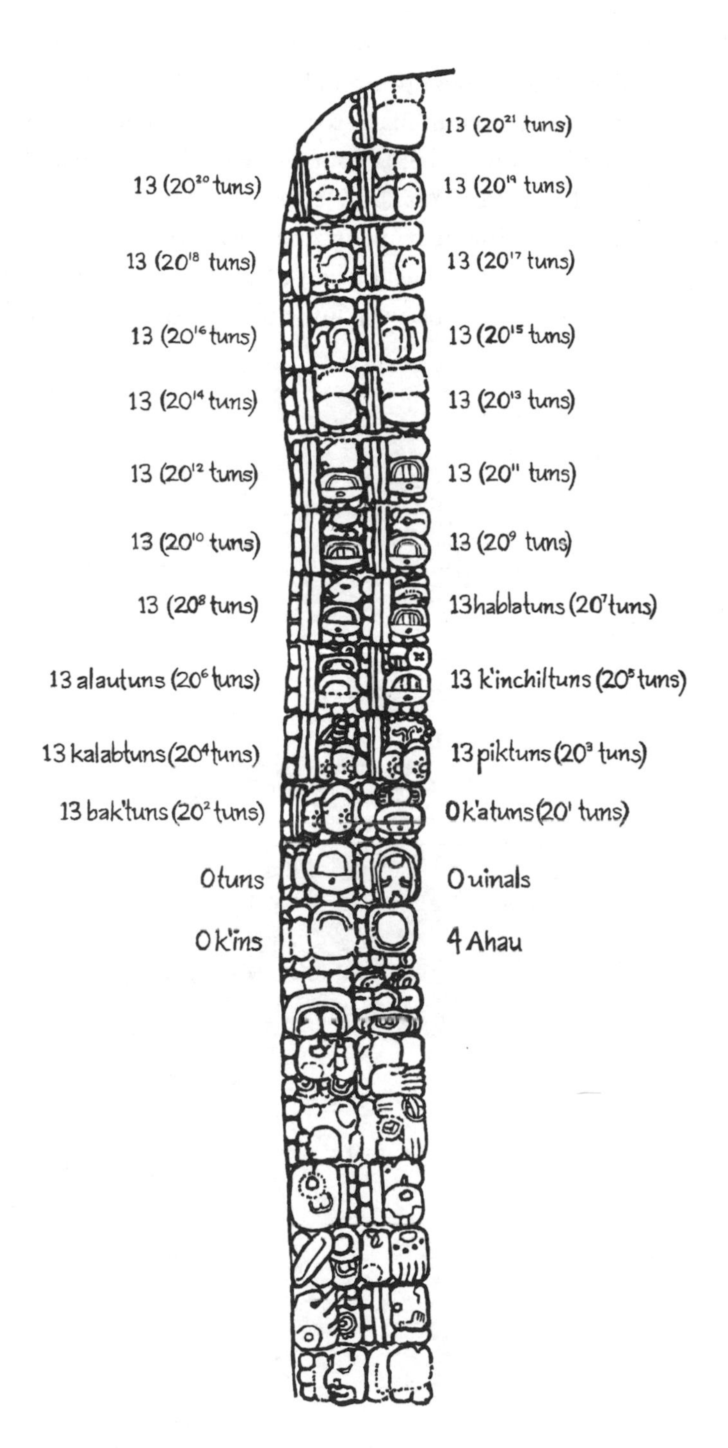

Fig. 2.2. *Coba stele. (Illustration adapted from Freidel, Schele, and Parker,* Maya Cosmos.*)*

We know a lot about what was happening during this Great Year, because civilizations suddenly appeared all over the planet at its outset. In this sense, that *is* when the human story began. Maybe that explains the fundamentalists' adamant belief that the world was created when the Hebrew Calendar began. Yet the Hebrew Calendar describes only the rise of human civilization, leaving out the great evolutionary struggle that preceded civilization. It leaves out all of the other species, the full story of creation, which may be why humans are so destructive of all other species today.

Fig. 2.3. *The Nine Underworlds of creation. (Illustration from Calleman,* The Mayan Calendar and the Transformation of Consciousness.*)*

THE RESONANCE OF THE TZOLKIN

What exactly are the essential aspects of the Great Year? (Do not forget that the Long Count and the Great Year are the same as the period Calleman calls the National Underworld.) To answer this, we need to understand that the 260-day Tzolkin (the day count still kept by contemporary Maya people) vibrates in sympathetic resonance with the Great Year by factors of thirteen and twenty, as already explained. We don't need to know a whole lot about the Tzolkin, which is a good thing, because the Tzolkin is very subtle. (To really feel the Tzolkin's power and to be motivated to handle the numbers and glyphs, you need to figure out your own day sign and contemplate its meaning, which you can accomplish by means of appendix D.) The Great Year and all its multiples is a tun-based Calendar, because tuns are multiplied by the power of twenty. Remember, each tun is 360 days long, thus tuns are not the same as the 365-day solar year; tuns track divine forces, not physical or astronomical cycles.

The Tzolkin, consisting of 260 days, is an everyday Calendar of the Maya, whereas the tun-based Calendar was created to understand the long cycles of evolution. The Tzolkin resonates with the Great Year because the Tzolkin is based on thirteen numerical factors, and its twenty glyphs are core archetypes that are multiplied by twenty. When something resonates with something else, it vibrates in tune with it, such as when a tone in each octave of a piano vibrates when that tone is struck in a single octave. The long cycles described by the tuns resonate with the twenty glyphs, which is the great mystery that confounds most people who attempt to penetrate the Calendar.

I want to emphasize the resonance factor because this concept is very important to the contemporary Maya, yet I still do not feel it very much day by day. As for this ability to resonate, the Ninth Underworld, the Universal Underworld, is only 260 days long, just like the Tzolkin. That is, this last Underworld has all nine Underworlds loaded into it day by day, by the 360/260 resonance. I don't think it is possible to comprehend such a fantastic idea except by imagining how the Maya might have discovered it. So I will use my own imagination for a moment. The Tzolkin is probably much older than the Great Year, possibly as much as three thousand years older, so I will take artistic liberty in what I am about to say.

HOW THE ANCIENT MAYA DISCOVERED THE CALENDAR

Going back into the mists of time when the people domesticated corn, the 365-day solar year or Haab was essential for farming. The people gave up their hunting-and-gathering way of life to become horticulturalists, and the cycles of the Sun became very important to them. When the people began to settle down on the land, they did not want to forget the brilliant shamanic skills they'd developed while foraging and hunting animals. After all, they had learned how to travel into the lower, middle, and upper worlds for knowledge, the worlds accessible by the Sacred Tree, the magical tree that connects all the worlds. When they became planters, they developed special ceremonies to bridge the newly settled world with the old flying worlds. They discovered the Tzolkin, a 13 × 20 divination system that keeps the count of the days; this kept them in contact with the spirit worlds.

Now that they were settled, they wondered how their villages connected with the sky, so they developed astronomy and sacred seasonal rituals. The people kept the count of days, which was a sacred gift from their ancestors. More and more, the people wondered where they came from and how long ago. They made special pots and placed them on their altars in deep limestone caves, and their ancestors came to dwell in the pots. The people wondered about the red ochre images of great reptiles drawn on the walls in the caves. They knew they were the people of maize because their ancestors told them they had co-created corn with the gods.

They made pilgrimages to the sacred mountain shrine where the original grain—teosinte—still grew, and they marveled at their ancestors' ability to coax this hard grain into maize. The sky reflected their lives on Earth, so they searched the skies for a center point. Knowing this center made it possible to safely travel anywhere in the sky, into the Upperworld, much higher than the top of the Sacred Tree.

Once upon a time, the shamans did something amazing. The maize people, who developed corn with the Creator and faithfully kept the count of days, found waves and cycles of daily creation in the Tzolkin. This is why they were the maize people! It takes 260 days to conceive and give birth to a child, so the count of days was like creating their own children, and they were the children of the gods. The two sacred numbers, thirteen and twenty, must contain secret codes, and so they played with them. The ancient ancestors had told them the sacred year is 360 days long. One day somebody thought, What if 360 is a unit, like the 365-day Haab is one solar year? This unit must be sacred, and so they gave it a name, the tun, which was exactly 18 times 20 days long. The solar year was their everyday life and work, while 360 was divine time. Seeking to know the time of their beginning, why not use the tun to create longer cycles? So they multiplied tuns by twenty, just like the twenty day signs they'd been counting for thousands of years.

In the 260-day Tzolkin, on some days the signs or glyphs told them that a particular day could be used for healing, while other days had many problems. Longer periods of time must be like that too. Just as the glyphs directed their days, maybe the Calendar would help them figure out when and where to build temples, or when to choose to live simply in the villages in their houses oriented to the sky.

Thus they multiplied tuns by twenty, which created katuns, which were almost twenty years long, the span of a generation. Just like the changes by thirteen numbers in the Tzolkin, they divided the katun into thirteen stages of creation, the Thirteen Heavens. Miraculously, these thirteen stages described the developmental changes in their civilization over twenty years; thus could they anticipate what would come next in their society! They were ready to build exquisite cities with beautiful temples because now they knew when to create and when just to dream.

They were astounded by the discovery of this sacred calendar because it must be how the gods create. To help their descendants remember this knowledge, they inscribed books, stele, and walls to record the cycles of the Sacred Calendar. They were pleased that they could predict when to dedicate temples to the gods, and when to connect their kings and queens to sacred powers. They made things work by their intentions, and they were ecstatic almost all the time. Longer spans of time began to feel organic just the way a day did.

Following the discovery of the katun, they multiplied katuns by twenty to find the cycles that describe the beginning, growth, and demise of their cities. Subsequent to this, they divided this new larger number by thirteen to discover the thirteen baktuns of the sacred Long Count.

The development of cities and ritual kings over long periods was like the growth of sacred corn, the growth process over thirteen days that began when the New Moon was a sliver in the sky until it was full like a woman ready to give birth. This synchronicity was amazing to them. When they observed one whole baktun (about 394 years), they could see that each baktun described longer historical cycles. Since they understood the thirteen numbers, they could prophesy the events that were to come in the succeeding baktun.

Meanwhile, the astronomers had already discovered the end-point of their civilization, when the winter solstice Sun would cross their center, the Galactic Center, around two thousand years in the future. So they used the newly discovered 13 baktuns to set up a Calendar going back to their beginning time, 3115 BC, that would end in AD 2011. Having found such knowledge, they did not stop there. The ancestors had told them never to forget that before their time there were other worlds or "Suns." They constructed longer cycles by multiples of twenty going back deeper into the past. They stopped at nine cycles that would all end in 2011, because their ancestors had told them there were nine levels of development guided by Nine Bolontiku—this was Earth's story under the sky.

Each one of these cycles had thirteen stages of growth from seed to maturation. By their legends of great ages or Suns, they knew they once were the monkey people, and sacred monkeys were always part of their ceremonies. Before they were monkeys they were reptiles, and long before that they were stars. When all this was understood, one of their greatest teachers and prophets inscribed the stele at Coba that goes back many cycles before the cycle of hablatuns (16.4 billion years), yet they had no names for those times so long ago.

THE PLAN OF CREATION

It is possible that the Maya who discovered the Long Count, a truly astonishing innovation out of the Tzolkin, just stumbled on it. Yet it may have been a legacy due to the fact that five thousand years ago, as previously stated, most ancient cultures—such as the Egyptians and the Vedic Indus culture—used a 360-day calendar.[5] Since a 360-day calendar was ubiquitous five thousand years ago, the ancient Maya could have retained the memory of 360 as the basis of sacred time in their oral tradition.

Against amazing odds, many indigenous cultures have used their oral traditions to save archaic knowledge, especially in the Americas. Still, they were probably quite amazed to see that in growth cycles multiplied by twenty, a katun functions like a tun, and mystified that a baktun might function like a katun multiplied by twenty. Because they tracked the growth of their own civilization by these time cycles, they could have stumbled on the evolutionary time-acceleration factor—the plan of creation itself.

This must be why indigenous people have struggled to protect the day count for thousands of years, which is in resonance with the tun-based Calendar. I doubt they could see what we now know to be true from modern science, yet I may be underestimating them. The Aztec mythology of five great ages, the Five Suns, is all based on human emergence from animals. This knowledge comes from their ancestors, the Toltecs, who built the great city Teotihuacan, which has many painted murals that look to me like portrayals of evolutionary cycles. In other words, the Five Suns of the Aztec may be analogs of the first five Underworlds.

It is likely that because the Classic Maya understood the qualities of the cycles, they had a very clear sense of their future. In 1989, I was in Council with three hundred Maya Elders at Uxmal in the Yucatan. I got into a discussion with three tribal people, who I believe were either Tzotzil or Tzutuhil. I'm not sure which, because we were all struggling to understand each other with a mixture of Spanish, Nahuatl, and English, as well as employing a lot of eye contact and doing much gesticulating. The tribal people were determined to find out something about me, and I was surprised they were communicating with me at all, so we kept it up. We sounded like monkeys chattering. Tom was with me during this meeting, and at this point he climbed a tree to get some privacy. I laughed when the leading Elders chastised him for elevating himself above the Elders and told him to get out of the tree! I'd just gone through two weeks of intensive ceremonial work during which nobody talked to anybody, and now I welcomed the opportunity to talk. Tom, on the other hand, always loved the silence.

The tribal people wanted to know where I lived and what I did, which was working as a publisher in Santa Fe, New Mexico. Once they realized I was the publisher of two of the shamans attending the meeting, Hunbatz Men and Alberto Ruz Buenfil, they wanted to know if I knew what was going to happen to my son and me in the United States. Then they proceeded by intense energy, gestures, sounds, and words, to precipitate a movie right in front of my face—the total collapse of America! They gave me a vision because they were worried for me if I did not know what was coming, which in fact I did, as you will see in chapters 6 and 7. Everything is right on schedule, and I can assure you the Maya do know what the end of the Calendar means for the United States. They are indeed powerful visionaries; they can conjure future time. We will go into these levels of work later, but now let us return to Calleman's evolutionary theory.

ACCELERATION OF TIME BY TWENTY

Now that we know a little bit about Calleman's model in figure 2.3, it is time to consider some very radical ideas: *Previous creations became part of later creations, and this evolution is generated by cycles that are twenty times shorter than the previous Underworld. Since each cycle is twenty times shorter than the previous one, time is accelerating by a factor of twenty during the opening of each Underworld.*

If you are confused about multiples of twenty versus processes that evolve by thirteen stages, the tun is multiplied by the *powers of twenty* to get the length of each Underworld. Then each Underworld is divided into thirteen phases of growth called Days and Nights. What exactly are these Days and Nights, and the Nine Underworlds? Calleman says the Nine Underworlds are "sequentially activated crystalline structures in the earth's inner core."[6] Figure 2.4 shows how tuns are multiplied by the powers of twenty to get the dates for the Nine Underworlds, and it compares the Underworld dates with the scientific dating of initiating evolutionary phenomena. As you can easily see, a stele carved around 1,300 years ago, which has been standing mutely at Coba ever since, actually describes the scientific cycles of evolution agreed upon by science during the last fifty years! Of course, you will have to read Calleman yourself for the deep databank and multiple ways to look at this idea—for example, how humans can apprehend the crystalline structure in Earth's core.

Figure 2.5 shows the multiples of the 360-day tuns and the duration of each one in time. Tun multiples are always by twenty, yet the tun itself is 18 × 20, which is the basis of the 360-day sacred year. As you can see, the tun is derived by multiplying 18 days (kins) by 20 days (a uinal). Of course, 2 × 9 = 18, and we have already noted that nine is an ancestral number that became the Nine Underworlds; this means 2 × 9 is the resonant factor of the tun. Since nine dimensions is the key number in

Underworld	Spiritual Cosmic Time	Physical Earth Time	Initiating Phenomena	Scientific Dating of Initiating Phenomenon
Universal	13 x 20 kins	260 days	?	
Galactic	13 x 20^0 tuns	4,680 days (12.8 years)	?	
Planetary	13 x 20^1 tuns	256 years	Industrialism	AD 1769
National	13 x 20^2 tuns	5,125 years	Written language	3100 BC
Regional	13 x 20^3 tuns	102,000 years	Spoken language	100,000 BC
Tribal	13 x 20^4 tuns	2 million years	First humans	2 million years
Familial	13 x 20^5 tuns	41 million years	First primates	40 million years
Mammalian	13 x 20^6 tuns	820 million years	First animals	850 million years
Cellular	13 x 20^7 tuns	16.4 billion years	Matter; "Big Bang"	15–16 billion years

Fig. 2.4. *The durations of the Nine Underworlds. (Illustration from Calleman,* The Mayan Calendar and the Transformation of Consciousness.*)*

my own work, later in this book I will speculate more on what the Maya knew about the powers of nine. The only thing you really need to see before I play with these cycles is that each of the Nine Underworlds is associated with evolutionary phases or kinds of consciousness that are *frequencies of creation that are speeding up by factors of twenty, and that all nine culminate simultaneously.* That is, all nine frequencies of creation are going on at once, like Beethoven's Ninth Symphony, as if it is 20 × 20 × 20, and so on. Looking at figure 2.3 again, as we go up the pyramid, time speeds up, and as we go down the pyramid it slows down, thus evolution is accelerating exponentially. Right now, we are in the middle of the Eighth Underworld (the Galactic) that began in 1999, which is only *12.8 years long!* Different frames of consciousness *do not replace each other,* nor do they *add to each other,* and they are all

Mayan Name of Period	Spiritual Cosmic Time	Physical Earth Time
Kin	1 kin	1 day
Uinal	20 kins	20 days
Tun	1 tun	360 days
Katun	20 tuns	7,200 days or 19.7 years
Baktun	20^2 tuns	144,000 days or 394 years
Piktun	20^3 tuns	2,880,000 days or 7,900 years
Kalabtun	20^4 tuns	158,000 years
Kinchiltun	20^5 tuns	3.15 million years
Alautun	20^6 tuns	63.1 million years
Hablatun	20^7 tuns	1.26 billion years

Fig. 2.5. *Tun-based cycles. (Illustration from Calleman,* The Mayan Calendar and the Transformation of Consciousness.*)*

completed simultaneously. This is a very difficult and delicious idea that totally confounds linear time.

SIMULTANEOUS THREADS OF CREATION

A theory of the simultaneous arrival of the stages of evolution in 2011 is an extremely radical idea, yet this is exactly what the mathematics of the Mayan Calendar describes. The idea of continually interactive threads of creation that all end simultaneously may brilliantly describe how we humans have evolved from cells, rocks, animals, monkeys, and hominids, and it may explain why we are still evolving on all levels and going faster and faster now. We are going so fast that I find it essential to have an idea of how Maya time acceleration works. Otherwise, being alive is like being pinned back by centrifugal force in a spinning basket ride, or being thrown back in a roller coaster at an unsafe amusement park. That is how being on Earth feels lately!

The idea that all life forms are derived from each other in progression drives the fundamentalists nuts. Ironically, if the world was created only six thousand years ago—the basic span of the National Underworld—then humanity is nothing more than ego expressing itself as the gory patriarchy, a profound estrangement from creation itself. This is a nightmare to me and to most sensitive human beings.

Contemplation of the Nine Underworlds invites us to realize our origins in the Galaxy itself, to imagine our own birth being like the birth of a supernova. I bring up supernovas here because I've often thought the inventors of the Calendar must have traveled as shamans into the center of the Milky Way Galaxy for their information. After all, I do it regularly with students during Nine Dimensional Activations, in which they learn to travel in nine dimensions with their consciousness. As of 2003, science has decided there is a black hole in the center of the Milky Way. In a black hole, time is radically distorted—eaten up—and it feels like it is slowing down or speeding up exponentially upon entering a black hole. If you can understand the physics of black holes—objects that cosmologists say have to exist in order for the universe to exist at all—then it is easier to imagine the physics of exponential time. I believe we can enter black holes with our consciousness only—it would be impossible to enter them in the material state—and that the black hole in the center of the Milky Way generates the evolutionary time codes.

We will delve into such complex ideas later. Now we need more comprehension of Calleman's Nine Underworlds. However, this whole discussion cannot proceed without accepting one fundamental paradigm: *The cycles described by the Mayan Calendar dates are the same as the evolutionary cycles described by modern science.* In fact, when the dates differ slightly, the Maya dates were probably *more* accurate, especially with regard to the creation of the universe, the so-called Big Bang. For this discussion, we can't figure out what the Calendar *means* until we are certain of what it *describes:* The Mayan Calendar describes the evolutionary cycles of the universe.

Underworld	Duration	Level of Consciousness Phenomena Evolved Frame of Life
Universal (Ninth)	13 uinals	**Evolution of cosmic consciousness** No limiting thoughts, timelessness No organizing boundaries
Galactic (Eighth)	13 tuns	**Evolution of galactic consciousness** Transcending material framework of life, telepathy, living on light, genetic technology Organized in galaxies
Planetary (Seventh)	13 katuns	**Evolution of global consciousness** Materialism, industrialism, Americanism, democracy, republics, electrotechnology Organized in planets
National (Sixth)	13 baktuns	**Evolution of civilized consciousness** Written language, major construction, historical religions, science, fine art Organized in nations
Regional (Fifth)	13 pictuns	**Evolution of human consciousness** *Homo sapiens* with ability to make complex tools, spoken language, art, early religion Organized in regional cultures
Tribal (Fourth)	13 kalabtuns	**Evolution of hominid consciousness** Human beings *(Homo)* who make complex tools and have rudimentary oral communication Organized in tribes
Familial (Third)	13 kinchiltuns	**Evolution of anthropoid consciousness** Lemurs, monkeys, Australopithecans with the ability to walk upright and use tools Organized in families
Mammalian (Second)	13 alautuns	**Evolution of mammalian consciousness** Evolution of multicellular organisms, sexual polarity, a continental structure and plant kingdom to support higher life Higher mammals
Cellular (First)	13 hablatuns	**Evolution of cellular consciousness** Step-by-step evolution of the physical universe: galaxies, stars, and planets; evolution of chemical elements Higher cells

Fig. 2.6. *Phenomena developed during each one of the Nine Underworlds. (Illustration from Calleman,* The Mayan Calendar and the Transformation of Consciousness.*)*

THE DAYS AND NIGHTS OF THE MAMMALIAN UNDERWORLD

You may want to refer back to the illustrations preceding this text to follow along with what I am about to cover next. The First Underworld, "the Cellular Underworld" of thirteen hablatuns, is 16.4 billion years long. This must describe the creation of the universe, since cosmologists say the date that it began was around 15 billion years ago. Each hablatun, which is one phase of the thirteen stages of this creation, is 1.26 billion years long.

There is not much to say about the phases of the Cellular Underworld, since the solar system formed only 5 to 6 billion years ago during the Fifth Day of the Cellular Underworld. The Second Underworld, the Mammalian Underworld of thirteen alautuns of 63.1 million years, is when the first animals evolved, and we have much more data on this Underworld, which spans 820 million years. The first animals were multicellular sponges and algae. These animals then evolved into fishes, amphibians, reptiles, mammals, primates, and so on. Science says this process began 850 million years ago, yet the Mayan Calendar is most likely more accurate.

Alautun	Day	Beginning MYA (MYA = million years ago)	Classes of organisms (modern estimate)
0	1	820.3 MYA	First clusters of cells (850 MYA)
2	2	694.1 MYA	First symmetrical soft-tissue animals, Ediacaran Hills fauna (680)
4	3	567.9 MYA	Cambrian explosion: Trilobites, ammonites, molluscs (570)
6	4	441.7 MYA	Fishes (440)
8	5	315.5 MYA	Reptiles (300)
10	6	189.3 MYA	Mammals (190)
12	7	63.1 MYA	Placental mammals (65)

Fig. 2.7. *The development of multicellular animals during the Mammalian Underworld. (Illustration from Calleman,* Solving the Greatest Mystery of Our Time: The Mayan Calendar.*)*

To understand how biological evolution is tracked by the Calendar, first we need to know about the Days and Nights, which are wave movements that alternate between creations and integrations. In each Underworld, there are seven Days and six Nights, totaling thirteen; new creation occurs during the Days, and the integration of that growth occurs during the Nights. Notice in figure 2.7 that the major transitions of mammalian evolution all occurred just when a Day began, and then these steps were integrated during the Nights. For now, notice how close the Mayan Calendar dates for new creations during Days are to the scientific dating for the classes of organisms. This level of concordance is mind-boggling! But now I leave the Mammalian Underworld, just asking you to absorb the accuracy of the Maya in charting the dates of these long-span biological transitions that science has recently described so well. It's important to see this concordance clearly, since the Maya discovered an evolutionary time wave that is also recognized by science. Just think how critical this wave factor would be during the

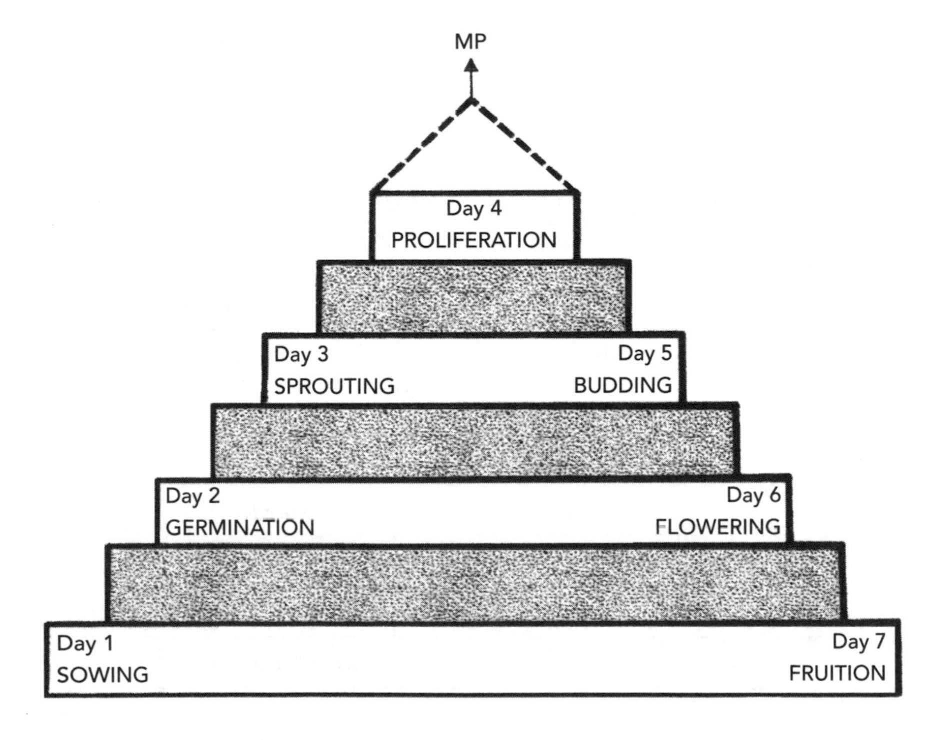

Fig. 2.8. *The Pyramid of the Days of the Thirteen Heavens. (Illustration adapted from Calleman,* The Mayan Calendar and the Transformation of Consciousness.*)*

shorter phases of evolution, such as the one we are in the middle of now, the 12.8-year Galactic Underworld. This is why humans so acutely feel the acceleration of time and the accompanying rapid changes. Aren't you happier if you have some kind of reason for the way things are going?

To discuss any of the Underworlds, we need to understand the cyclical functioning of the thirteen numerical divisions by Days and Nights, because each one of them describes the qualitative phases of the Nine Underworlds. As already mentioned, the Maya understanding of the Thirteen Heavens comes out of what they had learned about the thirteen numbers by working with the 260-day Tzolkin. These thirteen numbers progress from sowing the seed on the first day through maturation or fruition during the last day. Throughout Mesoamerica, there are pyramids with thirteen steps—six steps going up, a seventh step on the top, and then six steps going down. These pyramids encode the growth principle of thirteen, just as pyramids with nine levels, such as at Tikal, reflect the Nine Underworlds. It is easy to feel these divine principles in Mayan architecture. The pyramids teach people who interact with them—whether they are conscious of it or not.

As Mayan Elders, we divine and teach by using this architecture. Each level of the pyramid is ruled by a god or divine principle, just as each number is. By considering the nature of that god or goddess, we can ascertain the quality of any level. As for the nine-storied pyramid of Tikal, it teaches about time acceleration. In 1988, I spent the night in the top chamber of this pyramid. I feel I accessed the time-acceleration codes that night, which may be why I can recognize the acceleration processes of the Nine Underworlds.

Maya architectural principles are very important, because they demonstrate how the divine functions in the material realm; they show how numbers and glyphs have divine attributes. These superb examples of co-creation required leaders with high consciousness, geometrically informed artistry, and devoted laborers who loved their work. Generally speaking, the gods and goddesses that rule the even-numbered Heavens are more nurturing and feminine, and the gods and goddesses that rule the odd-numbered Heavens are more male and warlike. The first six numbers are the creation and building of an issue, the seventh is a creative explosion, and the last six add complexity that results in new creation.

Going back to the Mammalian Underworld when plants and animals evolved, at least several mass extinctions took place very close to alautun shifts, such as the extinction of the dinosaurs 65 million years ago. Right after this, placental animals appeared 63.1 million years ago, which is Day Seven of the Mammalian Underworld. During Day Seven of any Underworld, evolutionary processes culminate, and often mass extinctions occur just before this point. This same factor operates during all the Underworlds, yet phases that culminate during Day Seven are the ones most likely to be preceded by extinctions.

Also, when a whole new Underworld opens, there is a tendency for extinctions and great changes in the previous Underworld. For example, the Familial Underworld (41 million years long) is when the first monkeys evolved, the direct ancestors of hominids who appeared during the Tribal Underworld (2 million years long). *Australopithecus* appeared during Day Seven of the Familial Underworld, which culminated 3.15 million years ago. Then when the Tribal Underworld arrived 2 million years ago, *Homo habilis* just suddenly appeared.

When a new Underworld begins, higher forms of life organize with more complex brains to receive greater information from the universe. Although the evolution of the previous Underworld continues, *the new Underworld brings in big changes because the time acceleration is so intense*. These time-accelerated critical leaps are the shifts that most confound archaeologists and anthropologists, because they just seem to emerge out of thin air. To understand this process let's take a closer look at the Tribal Underworld of the last 2 million years, because it is the story of our human emergence.

THE TRIBAL UNDERWORLD AND *HOMO HABILIS*

The accompanying illustration is adapted from *Making Silent Stones Speak*, anthropologists Kathy Schick and Nicholas Toth's excellent story about the last 2 to 4 million years.

Here we see a great portrayal of what happened to two branches of *Australopithecus afarensis*. We learn that one branch produced *Homo habilis*, who arrived during Day One of the Tribal Undeworld (2 million years ago), and later evolved into *Homo sapiens* when the Regional Underworld began (102,000 years ago).

Next, we find that the second branch generated *Australopithecus aethiopicus,* a strain without cerebral reorganization of brain structure whose derivatives finally died out about a million years ago. *Homo habilis* preferred using the right hand, which probably encouraged tool making. "The result," write Schick and Toth, "was a more intelligent, foresighted creature with increased behavioral complexities able to transmit more information through learning. For the first time in the evolution of life on the earth, a *complex feedback loop between culture and biology began to emerge* [italics mine]."[7]

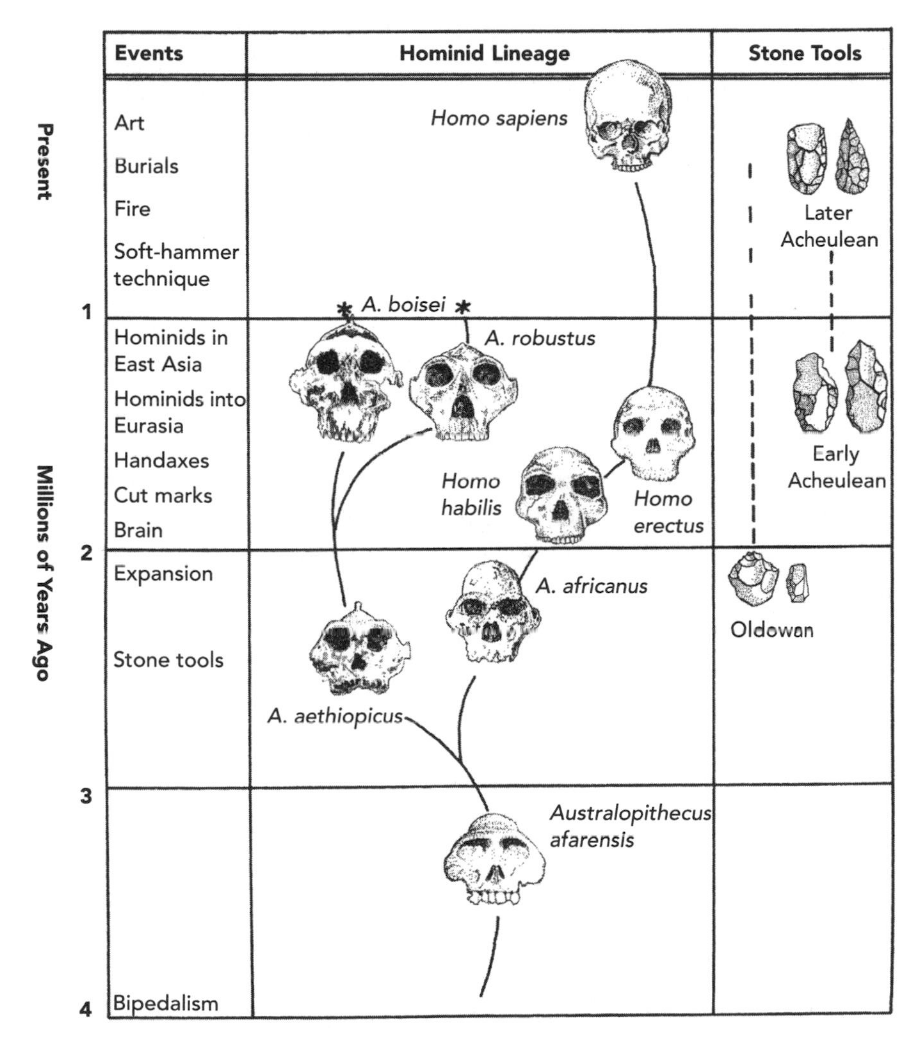

Fig. 2.9. *Possible evolutionary paths of early hominids to early humans. (Illustration adapted from Schick and Toth,* Making Silent Stones Speak.*)*

Authors Schick and Toth do not seem to know anything about the Mayan Calendar, and thus they express intense wonder at how cranial capacity increased and brain lateralization "suddenly" appeared exactly 2 million years ago when *Homo erectus* adopted tools. The authors trace the development of tool technology, which was at first quite crude, and then becomes more refined during the rest of the Tribal Underworld. *Homo habilis* evolved into *Homo erectus* about 1.7 million years ago, which was Day Two of the Tribal Underworld. Evidence from Olduvai Gorge shows that during the same period, *Homo erectus* accelerated their evolutionary path by adopting tools that emphasize "thought, cultural sharing of information, and planning for a future more removed from their immediate present."[8] This technology evolved significantly during each subsequent Day of the Tribal Underworld.

Homo erectus began migrating to Eurasia, spreading their culture. Their tools evolved to the point of being aesthetic, indicating that they valued artistic beauty, which would suggest some level of symbolic communication or language. Schick and Toth comment that during these hominids' 2 million years of tool making, there was remarkably little change in the form of their tools, the Acheulean hand axes.

In light of *Homo erectus's* ability to spread their culture, Schick and Toth write, "This conservatism [of *Homo habilis* and *Homo erectus*] is absolutely astounding. We never see anything else like it in our recent prehistory or history."[9] "Recent prehistory or history" in this case applies to the last hundred thousand years. Of course, by now in this book, we know what happened. It took the Thirteen Heavens of the Tribal Underworld (2 million years) to inspire *Homo erectus* to evolve into *Homo sapiens,* who are very much like modern humans. Unlike the 2 million years it took to learn how to make tools, a phase that was protected by the endemic conservatism of previous hominids, everything suddenly sped up a hundred thousand years ago. This shift befuddles anthropologists, yet the principle of a periodic speedup by a factor of twenty in the Mayan Calendar perfectly explains these radical transitions.

THE REGIONAL UNDERWORLD AND *HOMO SAPIENS*

Sudden and dramatic changes occurred 102,000 years ago: tools became smaller and more sophisticated with evidence of much artistry, and *Homo sapiens* began burying their dead. Then during the Regional Underworld around forty thousand years ago, there was an abruptly radical and creative innovation in all aspects of hominid (what is by now human) life. Then about ten thousand years ago, many groups adopted agriculture—the Neolithic Revolution.

Homo sapiens rapidly developed an advanced modern anatomy, with the potential for much greater intelligence than prevailing hominid species. Because *Homo sapiens* was an anatomically modern human, Schick and Toth see a great gap between the potentially sophisticated culture that *Homo sapiens* could have created using their modern biology, versus the relatively primitive culture that they actually did create. As these authors put it, "Between 100,000 and 40,000 years ago, these modern hominids do not appear to have done anything dramatically different from other hominids during this period, including archaic forms of *Homo sapiens* such as the Neanderthals and earlier Middle Stone Age hominids from Africa."[10] Considering the Days and Nights of the Regional Underworld, 37,500 years ago is the opening of Day Five, the Day when major breakthroughs occur in the cycle. Day Five of any Underworld is when we can see the potential for the remainder of an Underworld. The past forty thousand years are usually referred to as the Late Upper Paleolithic, when humans began to exhibit their new potential by means of exquisite art in their ceremonial caves.

As you can see so far, Calleman's analysis of time by means of the Nine Underworlds adds more information to the human story than told by anthropology, biology, and archaeology. I have really enjoyed surveying modern scientific research and using it to mentally travel way back in time. As a result, I have a newfound respect for the anthropological study of the last two million years of hominid evolution. But in my opinion, many anthropologists and archaeologists do not have enough regard for humankind's more recent accomplishments, especially for those advances initiated by the arrival of the Regional Underworld 102,000 years ago.

I will end this chapter with that comment, because now I need to diverge from modern science and favor some new-paradigm sources. In the next chapter, we raise the curtain on the Regional Underworld. Based mostly on my own research published in 2001 in *Catastrophobia,* I will reveal that over the past hundred thousand years, humans developed a much more advanced culture than what is supposed by modern archaeology and anthropology. In addition, I will add new insights that have come from absorbing Calleman's theory of intelligent design in the Mayan Calendar.

3

THE GLOBAL MARITIME CIVILIZATION

THE NEOLITHIC AGE

According to academic anthropology and archaeology, the Neolithic Age began about ten thousand years ago, when humans first manipulated their environments by farming, domesticating animals, and settling in villages. This widely accepted interpretation assumes that humans did *not* manipulate their environments very much before this time, but this is not the case. In this chapter, I present some examples of major environmental manipulation by humans prior to ten thousand years ago.

Regarding the conventional timeline, academic researchers expect us to believe that around five thousand years ago, when the National Underworld opened (3115 BC), advanced civilization just "suddenly" arose all over the globe—seemingly out of thin air. Yet, these societies required an elite to oversee building plans, protection, and their economies, so where did the elite come from? Of course, in this book, I follow Calleman's research and theorize that the sudden shift around 3115 BC was triggered by the evolutionary acceleration factor of the National Underworld. Yet, even with this acceleration, the field in which civilization emerged must have already been fertile for building cities. In fact, regarding very advanced archaeological sites, such as Çatal Hüyük in what is now Turkey, from 7000 to 5000 BC, the ground was fertile for an advance. The radical rise of civilization in 3115 BC differs from such Middle Eastern sites because it was patriarchal and hierarchical, a significant shift from the previous matrilineal cultures that worshipped the Goddess, such as Çatal Hüyük. Civilizations with advanced architecture, politics, writing, accounting, and mythology don't just emerge from nothing. This outdated archaeological and anthropological premise doesn't make any sense, but nevertheless this inane dogma is heavily reinforced.

Following Calleman's premise, time speeding up by twenty times 5,125 years ago would have generated an enormous change in the slow and idyllic life that characterized the 102,000-year-long Regional Underworld. According to anthropology, hunters and gatherers opened the first chapters of human history around one hundred thousand years ago, when life was as close to paradise as humans have ever achieved.[1] They say this harmony began to break down when horticulture—people settling down with gardens—was adopted during the early Neolithic Age.

I will explore the possibility that some people adopted horticulture, even early agriculture, and possibly built maritime cities twenty thousand years ago, and that the previous harmony was lost as a result of weather and Earth changes that began over twelve thousand years ago. Earth changes are the reason archaeologists and anthropologists find few remnants of advanced human activity prior to ten thousand years ago. We don't find many vestiges of earlier farming and even earlier cities because a great cataclysm 11,500 years ago devastated most of the planet and destroyed the remnants of a global Paleolithic civilization. This is a lost world, which some call Atlantis, and recently many new-paradigm writers have been describing it as a global maritime civilization.[2]

Regardless of a more ancient lost world, when city cultures developed five thousand years ago, the idyllic hunter-gatherer and early horticultural life of the Regional Underworld disappeared. This loss or disconnect that lurks in our subconscious minds was caused by the evolutionary speedup of the National Underworld, as well as by residual trauma from the great cataclysm. This loss is remembered as the "Fall," the time when humanity was kicked out of the Garden of Eden.

THE REGIONAL UNDERWORLD AS EDEN

Stories about the Fall suggest that the emergence of temple-city cultures was a very painful experience for most people. The new elite patriarchy received many privileges in the process, while the vast majority longed for Eden. For nearly one hundred thousand years, women were the gatherers of plants, herbs, and mushrooms, and they were expert healers. *Women were the professors and doctors of Eden.* Some anthropologists believe the men hunted only three hours a week, which meant there were long hours for ceremony, art, and dreaming.[3]

In case you doubt this male-female work ratio, hunter-gatherer societies that exist today pursue the same way of life. Do not forget that the Regional Underworld is still evolving on Earth through 2011. Hunter-gatherer societies, such as the Pygmies and Bushmen of Africa, the Tiwi of Bathurst and the Melville Islands north of Australia, and the Jarawa of the Andaman Islands fight to keep their way of life, since it offers them so much time and pleasurable activity, and it can be much safer than civilization![4] After the 2004 Indonesian tsunami, rescuers wanted to see if the Jarawa tribal people of the Andaman Islands had survived. Just like the wild animals in Sri Lanka that fled for high ground just before the tsunami hit, the Jarawa people knew how to read the signals of nature. They fled into the hills, and most survived.[5]

In this chapter, we will seriously consider the possibility that in some ways the people of the Regional Underworld had a more balanced way of life than modern humans. Why? *Hunters and gatherers do not control their habitat, since they are part of it.*[6] Modern humans are radically separated from the powers of their environment. They've usually lost their spiritual powers, the powers that enable humans to work *with* nature. I've become convinced that deep inside our minds, many of us remember the idyllic and powerful feeling that dominated the Regional Underworld. Yet few can grasp its mentality because our world—our habitat—is perceived rationally. Modern humans are usually left-brain dominant, which blocks the signals from nature that are picked up by the right hemisphere of the brain. The Regional people had a science of human participation in nature that did not preclude technology, and which we can still access today. The only way we can return to the Garden is to fully perceive nature. As we explore some of these very new, yet very old ideas, remember what we said before: each Underworld is stacked on the previous ones, and they are all developing simultaneously. In the current Galactic Underworld, we are experiencing the most rapid time acceleration we've ever known. It is possible to decode this synchronistic churning whirlwind of time by considering the parallel phases of the other Underworlds.

The Regional Underworld continues to unfold today, and I concentrate on it in this chapter because I think most of us are still processing the speedup of the National Underworld within the Regional Garden of Eden five thousand years ago, when we first became like modern humans. Of course, that speedup intensified again in 1755 with the advent of the Planetary Underworld, and now we are struggling with the wild speedup of the Galactic Underworld since only 1999! Now it is time to consider a relatively recent event that gravely disturbed our understanding of time.

THE GREAT CATACLYSM IN 9500 BC

Only 11,500 years ago, a great cosmic cataclysm battered our planet, and this global trauma is stored in the deepest recesses of our minds and bodies. It occurred during a rich maturation phase of the long, slow Regional Underworld, and against fantastic odds, people all over the planet retained the story of the day Earth nearly died. Similar legends about the day Earth shook exist in almost all of the oldest cultures on our planet. During the past 256 years (Planetary Underworld), as scientists reviewed the geological and biological evidence, they realized they had uncovered massive physical evidence for a great destruction that radically altered our planet.

The emotional implications of remembering what must have happened were so overwhelming that scientists have been very slow to accept their own data. Once geologists and paleontologists had surveyed the layers of rocks, soil, and fossils, they were surprised to find time went back so far. After all, according to the theologians of their day, the universe was created only six thousand years ago. Next, many scientists theorized that evolution proceeded by slow change over millions of years, although there were some who argued there was evidence for periodic catastrophes.

By the 1960s, so much global evidence for periodic catastrophes had been found that some geologists and biologists theorized that evolution is "punctuated" by periodic disasters. During my entire life, I've watched science try to avoid looking at the recent and most important cataclysm, the Pleistocene Extinctions. Instead, they focus on extinctions way in the past, such as the demise of the dinosaurs 65 million years ago. Maybe this is because we know that bands of hunter-gatherers wandered Earth

11,500 years ago. That is, we resist knowing about the recent cataclysm because humans experienced the nearly complete annihilation of life on Earth. Evidence of this catastrophe lies in the intact frozen bodies of oxen, bison, horses, sheep, tigers, lions, hairy mammoths, and wooly rhinoceroses found interred in the Sibero-Alaskan permafrost.[7]

As mining activity increased in the eighteenth century, much proof of mass extinctions was unearthed, which literally shocked the modern human psyche during the Planetary Underworld acceleration. And now that the permafrost is melting for the first time since the cataclysm, more evidence of this terrible massive death is being found. How has all this recent pain and death damaged the human consciousness? Lately, the climate is changing in some worrisome ways, which makes people wonder more about climate change and Earth changes thousands of years ago. Increasingly, people wonder how our ancestors handled climate stress.

Meanwhile, the indigenous people saved the ancient legends, including the knowledge of how to survive on Earth. For example, as previously mentioned, the Jarawa culture is sixty thousand years old and derives from Africa, yet during the tsunami, they knew what to do![8] This discovery is so valuable because we rarely get a glimpse into how hunter-gatherer cultures survive: it took a tsunami to reveal these remarkable Jarawan abilities. Long ago, when the people began to gather again and reassemble their lives after the disaster in 9500 BC, the storytellers who remembered what had happened were treated like gods. The storytellers could see that the people believed the environment was hostile, especially the sky. The people searched the sky nightly to see if it would wreak destruction on them again, and around their campfires they told and retold the old story of when the Earth nearly died. "What happened to us? What could it have been?"

The monster came between Sirius and Regulus, through the Pleiades, into the solar system and approached Earth, and then the horrible nightmare began. Earth's atmosphere became electrically charged, and the waters and air heated up so quickly that the people fell down on their knees in terror. The sky writhed with fiery serpents and dragons, Earth tipped toward the chimera, and then there was a deafening explosion. The magnetosphere and the chimera intertwined, and there was a resounding crack! Within hours, blocks of ice, hail, and gigantic masses of water assailed the people.

Great electromagnetic storms overwhelmed the bioelectric fields of animals, humans, plants, and even rocks, which is why we still have such mental blocks about this event. On that day, fear was imprinted deep in the reptilian brain, and this memory has been changing our relationship with our environment ever since. Next, chaos reigned as volcanoes exploded and the oceans and lakes boiled; Earth shuddered and cracked insanely.[9]

The scientific evidence for the great cataclysm is covered extensively in my 2001 book *Catastrophobia*, and in D. S. Allan and J. B. Delair's 1997 book *Cataclysm! Compelling Evidence of a Cosmic Catastrophe in 9500 B.C.* The late D. S. Allan was a science historian, and J. B. Delair is a geologist and astronomer. They assisted me in describing the global nature of the cataclysm, so that I could explore the effects of this cataclysm on human consciousness.

As for the cause of the cataclysm, they've found evidence that fragments from the Vela supernova came into the solar system and struck the

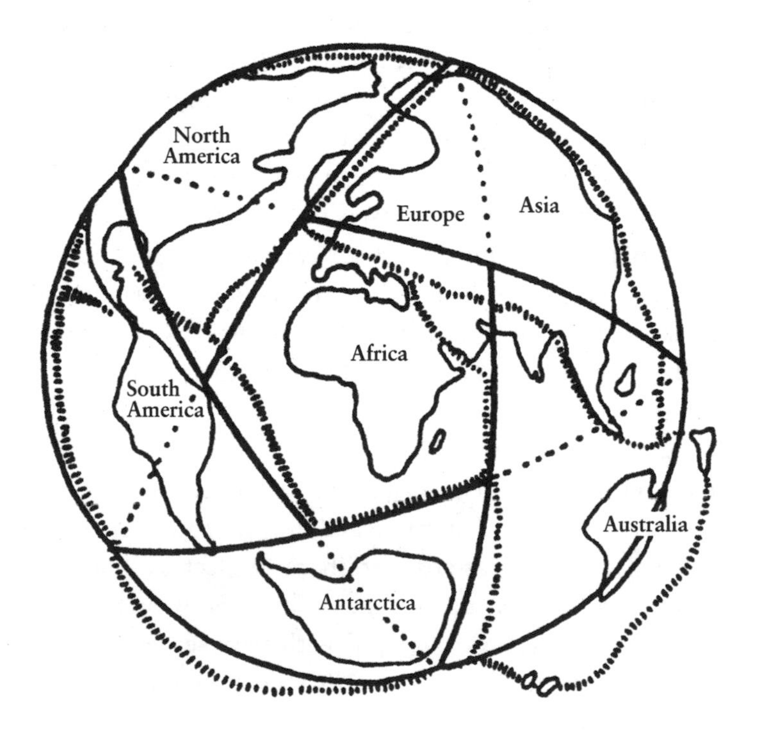

Fig. 3.1. *The icosahedral Earth. (Illustration from Allan and Delair,* Cataclysm! Compelling Evidence of a Cosmic Catastrophe in 9500 B.C.*)*

Earth's magnetosphere. It is difficult to know for sure what celestial body caused the event, yet it is certain it happened. The Earth's lithosphere was dislocated both vertically and horizontally, and tectonic plates formed that caused the planet to become a polyhedron of twenty faces—an *icosahedron*—as you can see in the accompanying illustration.

This seemingly outrageous and arcane detail is very important. The plates actually do have this form, yet it took Allan and Delair to notice it. This really caught my eye because in Cherokee tradition, Earth science is called "turtle medicine," and turtle shells have twenty plates. I'm sure this is an ancient memory of the shattering of our planet, and I heard an Iroquois medicine woman speak about turtle medicine in this way in 1986. During the Classic Period, First Father is often depicted emerging from the back of a turtle, which represents the three stars in Orion's belt that are the turtle's back.[10]

THE HABITAT AND POST-TRAUMATIC STRESS SYNDROME

Comprehending the magnitude of this event is crucially important because it implies that the most violent Earth changes are in the past, and that Earth is now settling down into a more harmonic pattern. Realizing that the worst is over and that Earth is rebalancing blunts the fear that the end of the Mayan Calendar means the destruction of our planet. Of course, this does not rule out more Earth changes, since Earth is still settling down.

As for whether humans will be destroyed in some form in 2012, this is up to us, since we are the ones who are making war on each other and on the environment. One reason for this book is to drive home the point that human mental and emotional blocks prevent remembering and processing the recent cataclysm, and they also are why humanity is so destructive to its own habitat. A cosmic disaster knocked our species out of the Garden, yet we still live on the planet. Because we struggle so hard to avoid processing our own pain, we tend to project an event that already happened into the near future, such as people do when they say the world is coming to an end in 2012. One of my contributions to Calendar research is to call attention to this insight. *Catastrophobia* posits that modern humans can't connect with this devastating event (and therefore remember it) because we are using an inaccurate timeline, which scrambles the data from the recent past. This idea is based on the premise that we have inner memory of the past deep within ourselves by means of racial, genetic, or past-life memory. Indigenous people protect the ancient stories of their own people because humans are less intelligent when they lose their memories of the past. Grandfather Hand, a carrier of the Cherokee records, insisted that the memory of the great cataclysm would be needed in our times. *Everything makes more sense if 9500 BC is adopted as a dichotomy point in human evolution.* The basic idea is that humanity was evolving through the 102,000-year Regional phase, a phase that was rudely interrupted 11,500 years ago when our planet was nearly destroyed. Being unable to understand what happened has been causing our consciousness to deteriorate ever since; the recovery of this memory is critical.

By recovering and acknowledging the correct timeline, I believe we can resume normal cranial evolution. What might that be? To survive, we need to reopen our right brains without losing the left-brain advances of the last five thousand years. We simply cannot evolve outside of nature; our home is Earth, not some other planet. This process could be a lot easier if people stopped resisting it, especially scientists. If you can open your mind, seeing the truth becomes a quest, a journey. For example, there is much evidence that soon after the aforementioned disaster, human knowledge was still more advanced in some ways than it has been until quite recently. Andrew Collins, Graham Hancock, myself, and others have written about an "Elder Culture" that existed eleven thousand years ago that instructed selected human societies to save ancient wisdom.

THE GLOBAL MARITIME CIVILIZATION

Stories about the Elders exist in Vedic, Egyptian, Native American, and Aymara records, as well as many more. The Elders assisted humanity for thousands of years after the cataclysm; they assimilated into the indigenous people, and we carry their blood. Evidence for this culture still remains on many continental shelves, which were above sea level twenty thousand to twelve thousand years ago.[11] Evidence for this global civilization has also been found in maps of the planet that go back thousands of years. Such maps were compiled by Charles Hapgood in *Maps of the Ancient Sea Kings,* in which he concluded that thirty maps that are

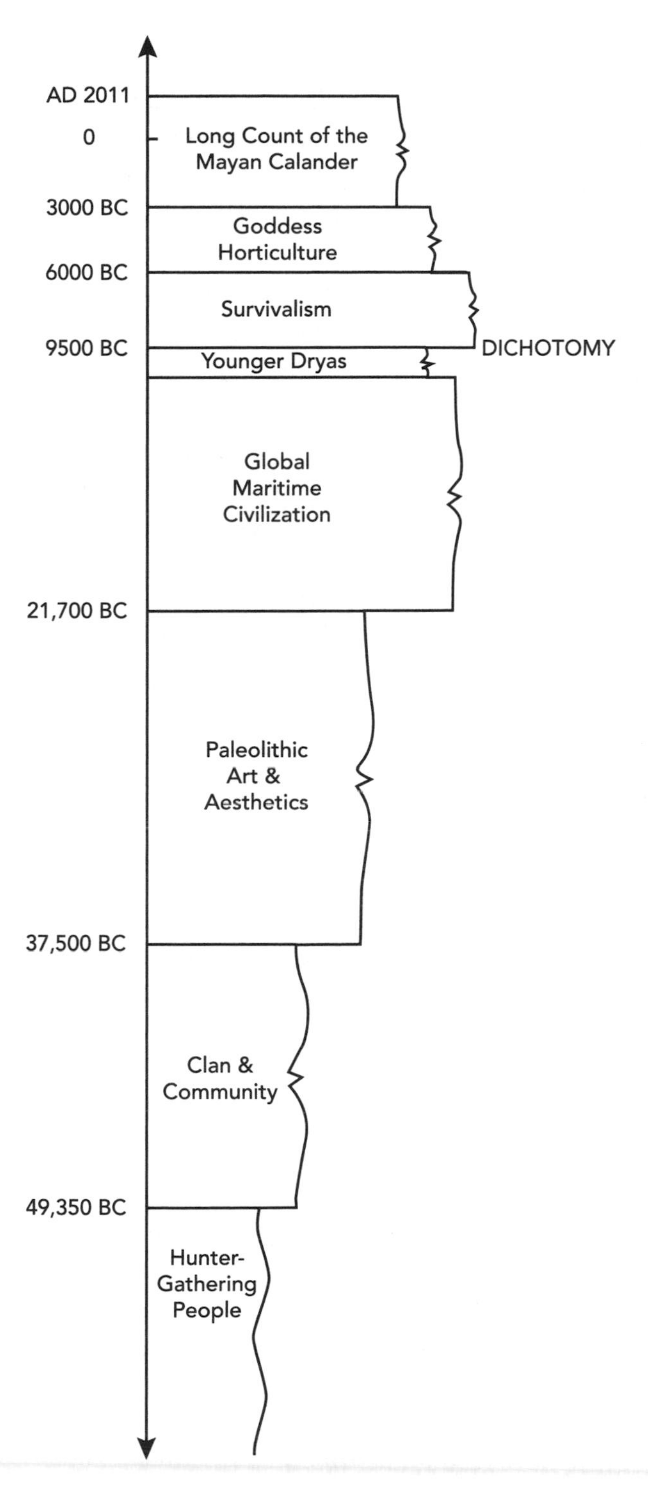

Fig. 3.2. *A proposed new timeline.*

thousands of years old were the tools of a scientifically advanced civilization that sailed the oceans more than six thousand years ago.[12]

Plato saved some historical accounts of global maritime cultures, such as Atlantis, Egypt, Greece, and the Magdalenian culture.[13] According to archaeology, from twenty thousand to twelve thousand years ago, the Magdalenian culture dominated southwestern Europe.[14] This evidence is already discussed at length in *Catastrophobia;* here I will mention some elements of this evidence that are relevant to this discussion.

Some new-paradigm writers are offering suggestions for new timelines based on the recognition of an advanced global maritime civilization in the past. But most of them do not pay enough attention to what the 9500 BC cataclysm did to the global civilization, creating conditions where survival was precarious and often worse than death. Even though these writers are sure (as am I) that humanity was once more advanced, they don't provide adequate explanations for what was then the massive regression of civilization, because they don't use 9500 BC as a dichotomy in the timeline. Without this divider, it is much more difficult to see what was happening: *We need to comprehend the regression of our own species, or humanity will commit ecocide.*

To straighten out the timeline, a few follow-up but localized cataclysms must be taken into account, because three subsequent major events are all mixed up in the records. There was a flood in 5600 BC—the Black Sea flood—that had a huge influence on human cultures, because it caused the dispersal of early advanced societies from the Black Sea region.[15] Forced from their villages by rapid flooding, the people left and carried their cultures to many places in the Middle East, eastern Europe, and eventually to Europe and the British Isles. Soon thereafter, complex megalithic sites were built, such as Carnac in Brittany. For these people, as well as those who experienced the huge flood of the Tigris and Euphrates—Noah's flood, around 4000 BC—localized floods revived the memory of the big cataclysm. These three events are all muddled into the story of the Flood in the Bible, and these complex threads are straightened out in *Catastrophobia*. I noted in the description of the cataclysm that human bioelectric fields were overwhelmed during the cataclysm, which affected our minds and bodies. I've noticed that when people get the correct timeline, these neurological systems reorganize. Many therapists, who have read *Catastrophobia* or attended my lectures, have reported to me that when they focus on these deep memory banks with clients, they find their clients become less fearful and paranoid. The cataclysm and survival period is a deep trauma-core in the human brain, and subsequent traumatic experiences appear to layer over and amplify this knot.

Meanwhile, as far as I can see, time acceleration is triggering a massive recovery from cataclysm-induced amnesia. During the Planetary Underworld (AD 1755–2011), this resulted in the recovery of Earth's scientific physical story, and now during the Galactic Underworld (1999–2011) people are searching for the true story of human cultures. People's obsessive curiosity about the hidden past is causing a quantum leap in human intelligence. Calleman's theory of evolutionary acceleration complements my proposed dichotomy in the timeline.

The melding of these two theories may represent a great potential to stop the regression and go forward. Also, the great cataclysm is a classic evolutionary extinction event that opened space for the next stage of evolution. It occurred just before the midpoint of Night Six of the Regional Underworld, when it nearly destroyed the global maritime civilization, remnants of which were still floating around after 9500 BC. This cleared some space for new evolution, which began when Day Seven of the Regional opened in 5900 BC. There is some evidence from 6000 to 3000 BC to support the view that enlightened Goddess cultures like Çatal Hüyük achieved remarkable progress after the disaster. This evidence is explored in detail in *Catastrophobia*.

What matters here is that humanity was *not* wiped out like the dinosaurs, but we did lose some brain access. I think a lot of this memory and much of the data from the earlier Underworlds is held in emotional trauma blocks and in what scientists call "junk DNA." According to science, only about 10–15 percent of human DNA is used for biological processes, and the rest is referred to as junk DNA. At this point in our evolution, given the converging time-acceleration factor of all nine Underworlds, our brains are reformulating rapidly. Assuming I may be correct that emotional trauma blocks are stored in the junk DNA (as well as memories of the past), it would seem that the speedups in time may be activating this DNA for use by our bodies and minds. If my thoughts seem presumptuous, I see much evidence for DNA activation in our students; sometimes they just light up with new awareness that is beyond what they seemingly have access to. In other words, the great knowledge we all hold within is waking up, and Gerry and I use healers when we teach because for most of us, waking up involves first breaking through trauma blocks.

As you will see in this book, the advanced Regional people had an astonishing relationship with nature. What if the DNA of the Regional people was fully activated, and what if that is required for human attunement with nature? What if these powers are coming back? After all, we are now in Day Seven of the Regional Underworld: the fruition phase.

Now that we have survived and covered the planet with our species again, it is time to assess and integrate the accomplishments of the lost global maritime civilization, the advanced phase of the Regional Underworld. We must understand our decline and the memory loss associated with it. Maybe the things we had to do to survive after losing a beautiful life have really injured our minds and hearts. Is this why humans are so hideously violent and act like desperate dogs that bite?

Fundamentalists are often very aggressive, hostile, and in denial. Maybe they can't handle the thought that humans once ran around in packs, scavenging for dead animals, and were sometimes driven to cannibalism? Yet, since this must be true, these horrific experiences are lurking deep in the human mind and soul. Dark and denied places in the human mind became visible to me when I spent some time with the Tana Torajans of Sulawesi, Indonesia. Our family was invited to a funeral, which turned out to be a bull-slaying ceremony. Out of respect, since we had already gifted tobacco to the relative who died, we had to sit for hours while live bulls were being butchered and carved up all around us. After enduring this ritual, in the evening our son Chris became violently ill, and its effect on our daughter Liz was such that she became a vegetarian. This experience helped me see that elements of archaic cultures can be very oppressive. During the bull-slaying ritual, I gave up a lot of my overly romantic ideas about the past being better than the present or future.

Most Torajans were hunters and gatherers only fifty years ago when fundamentalist Christians, who were determined to end the ancient practices, began to infiltrate their culture. The Torajan megalithic stone circles and graves carved out of huge rocks are some of the finest on Earth, and in the early years of the twentieth century the Torajans still knew how to erect standing stones weighing many tons. Yet, by the 1970s, when it was time to erect a stone for the installation of the next king of Toraja, they'd forgotten this art. They brought in a huge crane, but they could not erect the stone, and it still lies on its side in the middle of the royal stone circle. They'd forgotten how they erected all the other stones, and—perhaps significantly—this king for whom they were trying to erect the stone became the last king of Toraja.[16]

Meanwhile, being able to believe that people did raise stones that weigh many tons opens one's mind to the possibility that people in the past were capable of doing things we just can't do now. In Tana Toraja in 1997, I discovered that atavistic cultural activities, such as their funereal bull rituals, are entry points into the trauma banks of the human mind. I was literally driven to write *Catastrophobia* after what we witnessed in Tana Toraja.

The ritual we observed is the same as the bull-slaying ritual of Atlantis that Plato describes as having taken place twelve thousand years ago![17] Vestiges of this ritual still occur in the bull rings in Spain every year. Tana Toraja is a major ancient sacred site. It is now part of Wallacea, an area of Indonesia separated from Sundaland by a line drawn by the nineteenth-century naturalist, Alfred Russell Wallace. Wallace drew this line to separate Sulawesi (where Toraja is located) from Sundaland, the Indonesian continent, most of which went under the sea eight thousand years ago. Toraja was not affected by this submersion, and its culture, which claims to be from the Pleiades, is one of the most ancient on Earth.[18] If the remains of Sundaland in Tana Toraja are any indication, it must have been an incredible culture with lively rites from archaic times. The great 2004 earthquake and tsunami off the coast of Sumatra awakened an ancient memory of Sundaland, which is essentially a lost continent, except for the remaining large islands such as Sumatra, Sulawesi, Borneo, and Java.

Grandfather Hand insisted that the story of Earth's shattering is the real key to healing our species now. He had a Rand McNally globe of the world with a light inside it, and in the evening, we used to spend hours contemplating how Earth used to be before the terrible time. When I first saw Allan and Delair's reconstruction of the antediluvian world in 1996, I could barely breathe, because it complemented what I was taught when I was so young; it helped me remember many things Grandfather Hand had taught me. My school education was painful for me due to the fact that I heard nothing there about the real story of time that had been imprinted in my mind. Going to school caused me to forget the story of time temporarily. Somehow, Grandfather knew the true story would

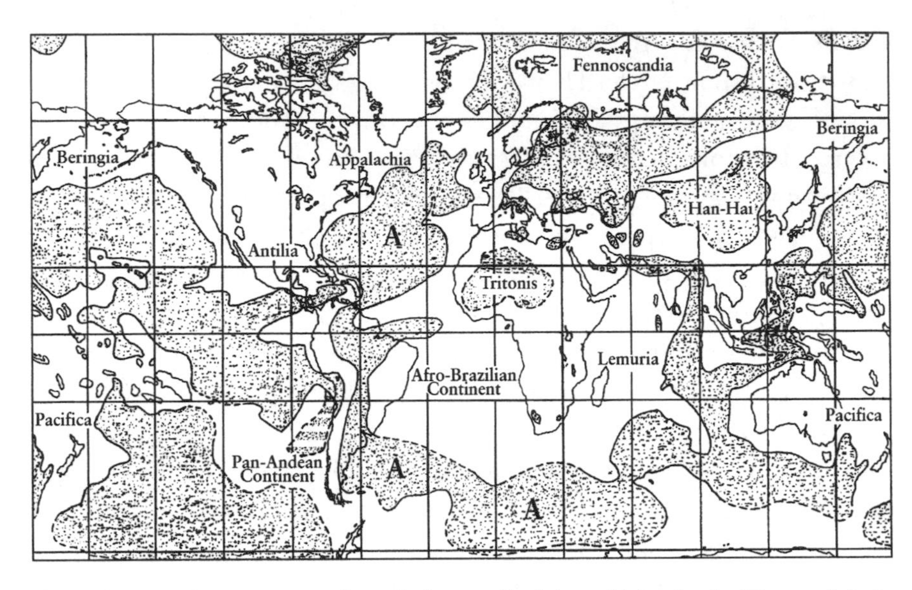

Fig. 3.3. *Tentative reconstruction of the prediluvial world by D. S. Allan and J. B. Delair. (Illustration from Allan and Delair,* Cataclysm! Compelling Evidence of a Cosmic Catastrophe in 9500 B.C.*)*

come forth during my lifetime because science would finally catch up with the indigenous records. As far as I know, he had not retained the end-date of the Mayan Calendar. Yet he somehow knew that he had to give the story that had been preserved by the Cherokee record keepers to his granddaughter, instead of to one of his five children—such as my father, who died in 1982. Also, Grandfather was a high-degree Mason, and I've often wondered if that is why he knew so much about the past and that an awakening would occur during my lifetime. In any event, the Maya/Cherokee records have much in common, yet the Maya saved the Calendar.

What is really going on as we wake up during the end of the Calendar is truly astonishing, and it will take this whole book to reveal this beautiful vision in time. In the meantime, I will briefly offer one of my more radical suppositions: Earth's plate tectonics demonstrate that our planet is an icosahedral sphere floating in space, which according to Allan and Delair, is a very recent form.[19] Icosahedrons are one of the five Platonic solids, the geometrical shapes that are the basis of how matter is formulated. In other words, *Earth transmuted into sacred geometry 11,500 years ago*. During the Regional Underworld, humans were enlightened because they were one with nature. Yet more acceleration in evolution was to come. The cataclysm may not have been a random accident, but a trigger to begin a higher harmonic state. Maybe becoming an icosahedral sphere was fundamental to Earth's attainment of enlightenment, and important to the whole Galaxy for that reason.

REMNANTS OF THE GLOBAL MARITIME CIVILIZATION

Little is known about the evolution of humanity during the Regional Underworld, since the cataclysm destroyed the evidence. I'm convinced that a few buildings from the global maritime civilization *did* survive, and they confound archaeologists and anthropologists. Examples of this are the Osireion of Abydos and the Valley Temple of Giza in Egypt, which are much older than nearby monuments from five or six thousand years ago, such as the Great Pyramid.

Both of these nonincised, cyclopean temples—the Osireion of Abydos and the Valley Temple of Giza—are located near or within dynastic archaeological sites. Archaeologists date them by the more recent ruins, yet both temples were obviously constructed by much earlier people; there is no good reason to lump them in with the surrounding early dynastic temples. According to some researchers, the layers of soil around them go back at least twelve thousand years, and even the respected geologist

Fig. 3.1. The Osireion of Abydos, Egypt.

Robert Schoch gave them much older dates than conventional archaeologists did.[20] They are thousands of years older than the dynastic sites.

My son Chris drew these figures to help readers see that cultures from before the cataclysm were aesthetically and technologically advanced; more than six thousand to twelve thousand years ago, people built architecturally exquisite temples by sculpting and placing stones weighing

Fig. 3.5. *The Valley Temple of the Giza Plateau, Egypt.*

many tons. How did they do this? And how did the previously mentioned Torajans erect huge stones less than a hundred years ago?

The Osireion is made of cyclopean stones that weigh many tons, and some researchers say it was built *more than* twelve thousand years ago.[21] In *Catastrophobia,* I presented evidence that the Valley Temple and the Osireion were buried in Nile silt during the cataclysm, and they were rediscovered and restored by the dynastic Egyptians.[22] Aware of the previous civilization by the Nile, these dynastic Egyptians restored the Valley Temple and the Osireion to remind people of their ancestors. The dynastic Egyptians were the proud inheritors of the culture of the global maritime Elders, the Shem-su Hor, and they revered these remnants from that prior civilization. Well, if there is even *one* building on Earth that was created by an advanced civilization more than twelve thousand years ago, then the past forty thousand years need to be reinterpreted! If there is even one map of the world that depicts Earth before the cataclysm, then geology and geography must be reinterpreted. In fact, there are such maps, such as the Piri Reis map that Charles Hapgood analyzed in *Maps of the Ancient Sea Kings*. As well, there are the enigmatic Stones of Ica, which I discussed in depth and Chris illustrated in *Catastrophobia.*[23] Briefly, the stones of Ica are a cache of engraved stones that were found in 1961 near the Nasca Plain in Peru. The stones were found in a cave—by the side of the Ica River—that had been exposed during a flood, and they are much more than twenty thousand years old. Some of them are engraved with maps of Earth from space, as well as humans coexisting with dinosaurs!

When the Pharaoh Seti I began construction of Abydos Temple around 1200 BC, the Osireion was discovered fifty feet below the level of the new temple he'd designed. Creating one of the best examples of excellent historic preservation, Seti restored the ancient temple and connected it to the new temple that has seven very unusual rooms in a line. These seven rooms, evidence for very advanced ancient knowledge, are the place where Abd'El Hakim Awyan instructed me about the "principle of seven." This experience helped me understand the seven-storied pyramid, the basis of the Thirteen Heavens of the Mayan Calendar, because these rooms represent the Seven Days of each Underworld. Hakim's legacy is at least 50,000 years old, and he has made it possible for me to penetrate the Regional Underworld.

DECODING THE REGIONAL UNDERWORLD

The Days and Nights of the Thirteen Heavens can be used as a tool to decode the Regional Underworld, which began 102,000 years ago. Looking at things in this way, we can see that the Regional Underworld was a time of very advanced creativity for humanity. The midpoint of Day Four, about fifty thousand years ago, was when the Paleolithic people must have deeply penetrated their habitat, consolidated their culture, and prepared to become global. Then the big breakthrough in symbolic art happened during Day Five of the Regional Underworld, during the middle Paleolithic period right after 40,000 BC, when there was a creative explosion that befuddles archaeologists and anthropologists. Anthropologists Schick and Toth note that this was when "fully modern symbolic expression" emerged.[24]

To begin with, what was this new culture that emerged when the period of the Regional Underworld began 102,000 years ago? We know from anthropology that suddenly around one hundred thousand years ago, bands of roaming people formed into clans, and the men and women delegated labor.[25] Because of this new level of communication, they must have discovered love, ethics, loyalty, and creativity. After

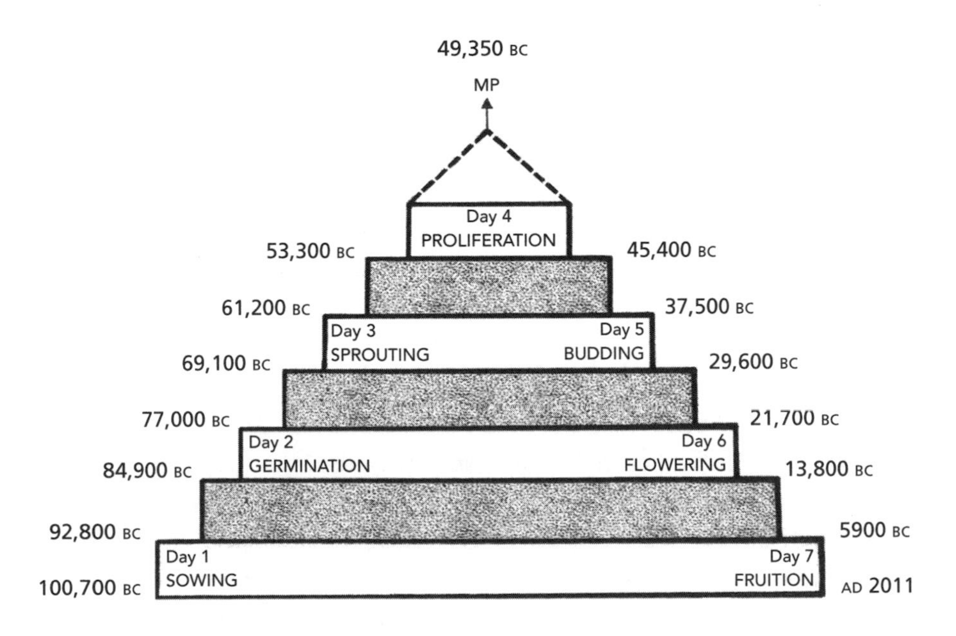

Fig. 3.6. *The Days of the Regional Underworld. (Illustration adapted from Calleman,* The Mayan Calendar and the Transformation of Consciousness.*)*

fifty thousand years of developing this new and very enjoyable way of life, new relations were natural with all the other clans. Each clan had adopted special art forms and beliefs, and now it was time to trade with each other to learn and share the wisdom of all peoples. These curious clans wanted to find other clans, so they built large boats to travel great distances searching for other sacred places and new people. According to Calleman, the major breakthroughs of any Underworld occur during the Fifth Day, which, in the case of this underworld, was 37,500 years ago. Thus, if you consider the "budding phase" of Day Five in the accompanying illustration, art and aesthetics came forth in the exquisite painted Paleolithic caves and the Goddess figurines, which suggest that in those days women were greatly respected because they could create life itself.

Anthropologists are puzzled by this aesthetic breakthrough because they underestimate the levels of culture that were actually being achieved at that time. *Catastrophobia* posits that the caves and rare early layers of ancient habitations are mostly what survived the cataclysm, while the infinitely more sophisticated maritime cities were completely destroyed and inundated by the rising seas.

I am pretty much in agreement with what archaeology and anthropology theorize about human evolution—up until the point of this aesthetic breakthrough in approximately 37,500 BC.[26] Thereafter, I think conventional science really misinterprets what human cultures were doing on Earth. Of course, the Regional people did not create and develop the way we have during the National and Planetary Underworlds. If you factor in cyclopean and megalithic stone technology, it seems obvious to me that the *Regional people had technologies that used nature's power*.

There is a modern mental barrier to understanding their technology, because modern left-brained humans separate and make distinctions between work and inventions on the one hand and nature on the other. The people of the Regional Underworld created reality first with thought, and then they manifested things in the solid world. They did not create anything that they thought would harm their habitat. They knew that if they did, it wouldn't work anyway, because then they would lose the power of nature.[27]

A new faith—matter over spirit—was born during the National Underworld and was perfected during the Planetary Underworld. This is why it is so difficult for modern people to imagine what the Paleolithic people were doing, much less feel respect for them. The old people knew that nature always reclaims her power; modern humans will eventually face a balancing, especially when the oil runs out. Many new ideas about what ancient people were up to are coming forth during the rapid acceleration of the Galactic Underworld.

UNITY AND DUALITY CONSCIOUSNESS

Why are we moderns so radically separate from nature? Calleman's concept of the Planetary Round of Light is an interesting attempt to answer this question. He thinks that the nine Underworlds each have a specific dark/light or yin/yang polarity designed to carry humanity on the path to enlightenment.[28] Each one of the Underworlds favors a specific hemisphere in the human brain, and these shifting lenses of brain access are influenced by Earth's geography and the location of the World Tree, the primary driver of the time-acceleration dynamic according to Calleman.[29] To quote him, "Because the brains and minds of human beings are in holographic resonance with Earth, as the cosmic pyramid is ascended [nine Underworlds], their resulting frames of consciousness will be dominated by the corresponding yin/yang polarities."[30] This is a very complex and intriguing idea, and you must read Calleman to fully understand what he is saying. Later, I will cover how the World Tree and the shifting polarities influence Earth. What matters here is that this concept of yin/yang polarity is very ancient and cross-cultural, which supports Calleman's conclusions that it is a big issue for human perception and brain function.

The basic idea is that as the Underworlds progress through time, consciousness goes from a state of being unitized to a state of duality and back to being unitized, so that transformational processes are generated. Knowing about this factor allows us to go with the flow and better anticipate what might happen next; as time accelerates, this ability seems to be crucial. The National Underworld consciousness, which is currently the most intense influence for humanity, functions in total duality. This is why we have become so radically separate from nature during the last five thousand years, since duality favors excessive male dominance. During the 16.4-billion-year Cellular Underworld, the 102,000-year Regional Underworld, and the

260-day Universal Underworlds, *consciousness is unitary and whole-brained* with no separation between the divine cosmos (nature) and man or woman; humans are enlightened.[31] As we already know, we are currently in various phases of the first eight Underworlds, and the Universal Underworld is still to come in 2011. Yet, non-unitized aspects of consciousness, such as the dualistic, left-brained developments of the National Underworld, are all developing simultaneously. Here I focus on the three unitary and whole-brained Underworlds to see how enlightenment might manifest in a more complex way in people whose left brains are relatively more developed. That is, enlightenment for humans will be different for humans of the Universal Underworld than it was for people living during the Regional Underworld. In this way, we can never go back in time.

During the Cellular Underworld, cells are enlightened, which is why cellular healing is the most potent form for us now. For example, you can't win a battle against cancer, but you *can* align your emotions and thoughts with your cells in order to heal your whole body, which can then deal with the cancer. As we have already seen, during the Regional Underworld, nature and humans are enlightened co-creators, which is why the healing of nature is needed in order to handle the last acceleration times twenty in 2011. As wonderful as Regional life has been, we have still greater stages to attain.

During the 260-day Universal Underworld in 2011, unitized, whole-brained consciousness will return to Earth for the third time, while we are aligned with the Galaxy. This period arrives while we retain all the knowledge we've attained during the National and Planetary Underworlds; at this point we will seek to integrate this knowledge to achieve enlightenment in 2011. This "brain integration" will transpire in a period whose length of time is less than one year, as we experience another acceleration times twenty. To be absolutely clear, *the First, Fifth, and Ninth Underworlds are the times when all that exists on Earth is in full holographic resonance with the cosmos*. Fortunately, I've noticed that many people born after the 1970s are very attuned to unity consciousness, yet they struggle painfully to survive within the dying embers of our dualistic civilization.

MEGALITHIC AND CYCLOPEAN STONE TECHNOLOGY

In the illustration below, consider the size of the stones of this beautiful temple (measured against the figure of the Peruvian man for scale) and the sophisticated techniques required to position them, and given that, try to understand how it could be that this temple may have been built more than ten thousand years before the era of Classical Greece.

The people who built this temple must have discovered an advanced technology for moving great stones, possibly by vibrating the stones with sound to lighten them. I'd like to remind readers who want to dismiss these facts about stone technology that until they can show, by means of modern technology, how these people moved and set up the great stones, they have no argument. The size of stones these people lifted is sometimes incomprehensible. Some of the most fantastic examples that we've seen are called *dolmens*. Dolmens are made of one huge rock that rests on three supporting stones, which are also huge. Dolmens are all over the British Isles, western France, and Spain, and I've even found them in the United States.[32]

One day, Gerry and I were driving around Brittany, and we saw a sign that said, "Stones Hotel." This is the kind of thing we do not pass up, so we turned into the parking lot. In the distance, we could see an inviting old hotel way down a bucolic garden path. Then we looked by the side of the parking lot. A huge egglike dark stone, as large as a Greyhound bus, sat merrily on top of three stones that each weighed a few tons!

Fig. 3.7. *Sacsayhauman Temple in Peru.*

The big stone weighed three hundred tons or more! I have seen one as large as this by the right side of Connecticut Highway 6, just after leaving Rhode Island and entering Connecticut.[33] In my experience, most people look at these things and blank out what they are seeing. I have a left-brained friend in upstate New York who has a wonderful small dolmen (ten tons) in her back yard, and she can't even see it. The facts are that eight thousand years ago somebody lifted these stones as if they were light as a feather! Sometimes I feel like they did it to play a joke on modern humanity, and they succeeded!

The engineer Chris Dunn explored the nature of advanced technology in his book *The Giza Power Plant,* which theorizes that the Great Pyramid was once a power plant.[34] Dunn began by proving that the dynastic Egyptians were carving stone with power tools more than five thousand years ago, which renowned archaeologist Flinders Petrie also concluded a century ago. Archaeology didn't follow up on Petrie's discoveries, since it would have meant the dynastic Egyptians were technological, so it took

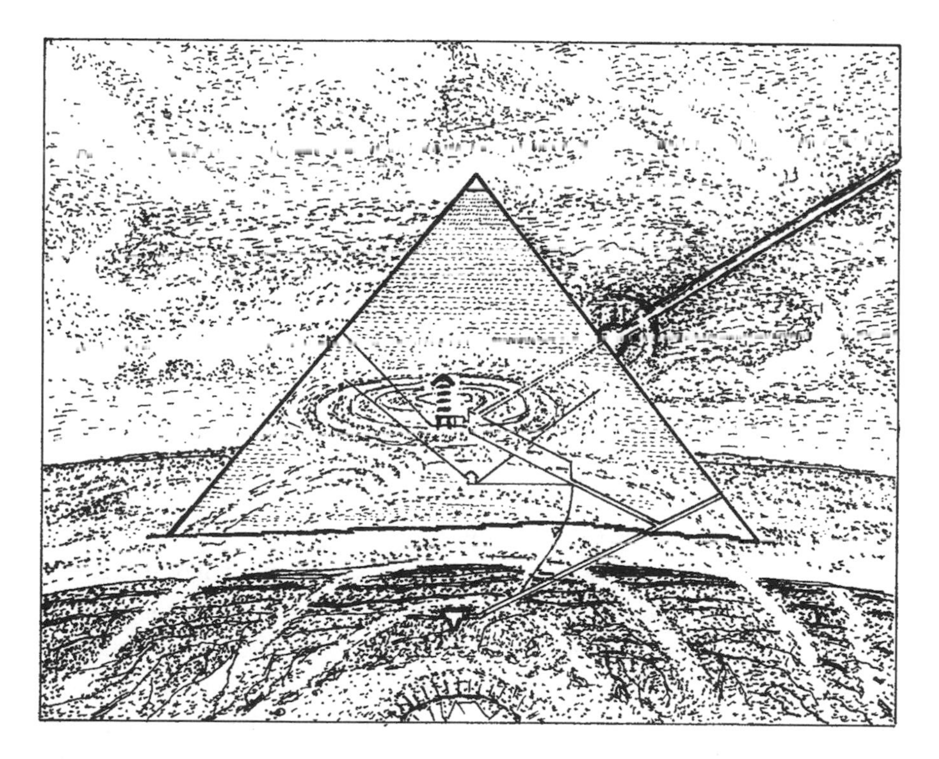

Fig. 3.8. *Coupled oscillator. (Illustration adapted from the cover of* The Giza Power Plant *by Christopher Dunn.)*

a modern engineer to do it. Next Dunn wondered about the source of power that ran those tools. He builds a strong case that the pyramid was a power plant built to tap Earth's energy and that it was constructed to be a coupled oscillator that can draw on Earth's energy once it is primed.[35] Figure 3.8 gives you a feel for this idea.

Next, Dunn needed some modern proof that this technology could indeed have existed way back then. He investigated the Florida laboratories of Edward Leedskalnin, who had claimed to know the secrets of the ancient Egyptian stone methods. Back in the 1930s, Leedskalnin built Coral Castle in Florida by lifting stones that weighed more than thirty tons and placing them in the walls of his castle.[36] Dunn concludes that Leedskalnin had built an antigravity device, which must be how the dolmens were lifted.[37]

I'd like to add that the early dynastic Egyptians did something else that is very striking. They built huge crypts and buried gigantic cedar sailing boats in them by the side of the Great Pyramid, by the side of the causeway of Saquarra, and as part of the early dynastic sites near Abydos. You can see one of them in a museum right next to the Great Pyramid. These boats are a reminder of the peoples of the global maritime civilization that sailed all over the world more than twelve thousand years ago, and who returned six thousand years ago to the sacred land by the Nile. These are the boats of the Elders who may have taught technological skills to the people, such as how to lift great stones and use resonant couplers and antigravity devices, which modern inventors are bringing back. I believe that dynastic Egyptians buried the boats by the Great Pyramid and at Abydos to remind their people and future people of the global maritime civilization.

DAY SIX OF THE REGIONAL UNDERWORLD

Using Calleman's seven-story pyramid in figure 3.6, let's put the global maritime civilization more in context. As we know, many new-paradigm researchers are convinced that a very advanced global maritime civilization existed all over the world from about twenty thousand to ten thousand years ago. The flowering of this global maritime civilization would have occurred mostly during Day Six of the Regional Underworld, from 21,770 to 13,800 BC, which is when I believe that the Osireion and the Valley Temple were built.

Once the Day Five breakthrough of Paleolithic art and aesthetics was attained during Day Five of the Regional Underworld (37,500–29,600 BC), people went to the next level during Day Six: They sailed the globe and built cities around their ports, and traded food and products with the people who lived nearby. Just as we like to do today, people built their cities by the sea, which made it easier to sail off to distant places for trade. They harvested the sea and grew their food in the rich alluvial soil in the river valleys that drained the mountains. For their religion, they must have visited the ancient ritual caves of their ancestors.

Since these global maritime cities and farming villages were destroyed during the cataclysm 11,500 years ago, the sacred caves are almost the only remains of their culture. This is why archaeologists genuinely believe the people were artistic but primitive. I think once they became sophisticated city people and sailors, they protected this visionary art, since it offers instructions for how to stay balanced with nature. For example, during the megalithic times, which are more accessible to us, all the spirals and circles etched in stone instruct about balance in nature.

As you will see later in this chapter, the ritual body postures discovered by Dr. Felicitas Goodman suggest there is continuity in culture from forty thousand to ten thousand years ago. However, for years after the cataclysm, the seas rose hundreds of feet as the shores of the continents rose and sank and the rivers and valleys silted and flooded. The ruins of the global maritime civilization lie under these seas, as well as in deep layers of silt in inundated river valleys all over the planet, which makes it very difficult to reconstruct this culture. This is why Seti I was so excited when he found the Osireion under fifty feet of Nile silt! These days, the artwork in Paleolithic caves is of great interest to Europeans because they sense a kinship with the artists who actually are their ancestors.

PALEOLITHIC CAVES

From twenty thousand to twelve thousand years ago, southwestern Europe was dominated by the Magdalenian culture, the main flowering of the Regional Underworld. Remnants of this culture have been found in caves on or near rivers that empty into the Atlantic in present-day Spain and France, such as the Lascaux Cave. Deep within these caves are exquisite paintings of bulls and horses, and many people have commented on their artistic sophistication and eerie beauty.[38] Evidence exists

Fig. 3.9. *Imaginary global maritime city below the entrance to Cosquer Cave, France.*

in some of this cave art that the Magdalenians bridled horses![39] As you will see later, one of the ritual postures discussed in this book was found in a painting in the Lascaux Cave.

In 1991, a diver found an undisturbed cave—Cosquer Cave—by discovering an entrance to it 137 feet below sea level on the coast of Mediterranean France.[40] The Cosquer Cave was used twenty-seven thousand through eighteen thousand years ago, when its entrance would have been located just above a descending plain that sloped down to a harbor. These days, this plain has become a continental shelf under the sea off the shore in this area. This would have been an ideal location for a global maritime port, just as nearby Marseilles is today.

Chris drew his own artistic rendition of a maritime city that may still lie in ruins on the extended continental shelf below the Cosquer Cave. In the illustration, note the large temple that resembles the Parthenon in Athens. Thousands of years ago, the entrances to the cave would have been just behind this temple. The sea level is now 137 feet above these entrances, and possibly three hundred feet above the imaginary city. Since this was a seafaring culture, when the cataclysm came, some people escaped in boats. (A similar escape from a natural disaster occurred in 2004, when the Sumatran tsunami swept people from the beach, while other people off Sri Lanka survived in boats.)

Plato understood the importance of the memories from before the cataclysm. In his day 2,500 years ago, only scattered stones and memory fragments of a lost world still existed, and even then many of his own contemporaries did not believe the story of the cataclysm was true! Thus, he wrote the story of Atlantis as real history, the source of all speculations about Atlantis. We would have many more sources for Atlantis if the Alexandrian Library in Egypt had not been partially burned by Julius Caesar in 48 BC, and later torched by Christian fanatics. Plato is the most important historical source for the global maritime civilization, yet academics refer to Plato's history as mythology.

The fact remains, however, that Plato's story of Atlantis is the oldest historical description of civilization from just before the cataclysm, and that is where historians ought to start. He describes a war between the Atlanteans, Greeks, and Egyptians, who were all sailing the Mediterranean and the Atlantic 11,500 years ago. Plato reports that Atlantis controlled the Magdalenian region, a detail that links cultures that flourished 30,000 to 11,500 years ago. This means the Magdalenian people were members of the global maritime civilization—as well as the Greeks, Egyptians, and Atlanteans.[41] Anything we have from this culture is a direct link to our Regional past, so let's look into it.

PALEOLITHIC LITHOPHONES

British archaeologist Paul Devereux recently made a truly awesome discovery about Paleolithic caves. He has shown that more than just visual art and ritual was going on in these caves; music was also being played in them. Devereux studied the acoustical properties of Paleolithic caves that were used forty thousand to twelve thousand years ago.

Key Paleolithic caves have various red ochre and black dots, lines, and symbols that many people have wondered about. Devereux discovered that these crude symbols mark the acoustical properties of stalactites and stalagmites, as if the stalactites and stalagmites are vaulted acousti-

Fig. 3.10. *Lithophones in the Paleolithic Cougnac Cave, France, with stalactites and stalagmites that appear to be acoustical pipes.*

cal pipes, similar to a pipe organ. The marks indicate which stalactites and stalagmites to strike to produce various tones within the caves. Devereux calls these stalactites and stalagmites *lithophones*.[42]

One of the most amazing ritual events of my life was being in a fire dance in Lol Tun Cave in the Yucatan while huge stalactites were being struck; they made delicious sounds that resonated in the cave cavity.[43] It would appear that Paleolithic people used their ritual caves as advanced sound devices or musical instruments—and this is still going on today.

Devereux also found much evidence for advanced acoustical science in manmade interior stone temples, such as megalithic cairns and passage graves. Devereux discovered that they were constructed to enhance the adult male vocal range, which would have intensified the effects of male chanting.[44] He was able to realize these insights because he started with the assumption that the original artists were intelligent. This information is covered in more detail in my 2004 book, *Alchemy of Nine Dimensions*.

THE MEGALITHIC GEOLOGICAL UNIVERSITY OF CARNAC IN FRANCE

Paleolithic people were using advanced sound technology, and there is much evidence that people after the cataclysm (during the megalithic phase) continued this science. A French engineer, Pierre Mereaux, studied the megalithic complexes of Carnac as well as those found around the Gulf of Morbihan in France's Brittany. These complexes go back seven thousand years and were in use until approximately four thousand years ago.

Fig 3.11. Carnac, France.

Although few people know about Mereaux's research, he has essentially proven that the people who built Carnac used it as a megalithic geological university where a priestly elite taught students about the power of sound waves and tectonic movement.[45] He concluded that these ancient people knew how to exploit magnetism to enhance human intelligence, genetics, reproduction, and healing powers, and he suggests that stones were erected because of their field effects on the human body.[46] I suggested in *Alchemy of Nine Dimensions* that the people of Carnac might have been using the healing technology of the megaliths to eliminate post-traumatic stress caused by the cataclysm and the rising seas.[47] Just as there may be an inundated city on the continental shelf off Marseilles, there may be one off the Gulf of Morbihan. Archaeologists have discovered many more stone circles, passage graves, and cairns on the shelf in the sea by the Gulf of Morbihan.

Gerry and I spent a lovely evening on the beach near Carnac watching the sun set behind a beautiful passage grave that is half under water on a little sinking island out in the gulf. We could have hired boats to take us to other sites that are being engulfed by the sea. Carnac has some exquisite cairns, and I hope Paul Devereux will be able to study their acoustical properties to see if the early builders of Carnac were using sound to heal. I've felt this possibility at many megalithic sites, which leads me to believe that an advanced acoustical and magnetic science may have existed in the global maritime civilization, and we only find diminished remnants of it at megalithic sites. Is it possible that we are very sensitive to magnetism because it tunes the brain?

I do know something about modern techniques for healing with sound. Gerry and I took John Beaulieu's Sound-and-Healing Seminar in 2000.[48] John Beaulieu is a great teacher of many healing techniques, and he developed the techniques for physical healing using tuning forks. At the seminar, we learned to harmonize frequencies in human organs by using calibrated tuning forks to adjust any organ frequencies that were out of balance. When I worked with the tuning forks over people's bodies, I could hear their organs humming at different pitches! These sounds were awesome.

Nearly instantaneous cellular healing can occur with this method, yet clients tend to avoid this form of healing because the correct vibrations unlock emotional blocks very quickly. For example, a really angry person can get a huge emotional release when their liver is vibrated, and it doesn't feel good at all because the release is so intense. It's my belief that the forks often release ancient cataclysmic trauma (as well as current lifetime blocks). From what we know about megalithic remains, it is likely that the Regional shamans used stones, caves, and cairns for healing and to help people stay balanced with nature, and now aspects of this technology are coming back.

The reason so few people know about these amazing findings is because when it comes to deducing the truth from ancient remnants, sometimes archaeologists are as prone to cranial rigor mortis as the fundamentalists are about the long cycles of evolution. Nevertheless, in spite of the prevailing and enforced dogma, new-paradigm researchers like Pierre Mereaux, Paul Devereux, and Graham Hancock have made fantastic advancements in data analysis during the last twenty years or so. Yet, many of them spend a lot of time trying to break down the orthodoxy, which just ignores them or debunks them. I think arguing about the old paradigm is a big waste of time now because time acceleration is waking people up, and the younger people are, the more amazed they are that such limiting ideas could have ever been in control in the first place. If archaeologists don't have the curiosity to consider Mereaux's brilliant ideas about Carnac, since they've never been able to figure out what the site was used for, why should we bother with their narrow-minded thinking? It is easier, more fun, and more productive to follow the new paradigm researchers, so let's consider one of my favorite teachers, Dr. Felicitas Goodman.

RITUAL BODY POSTURES AND ECSTATIC TRANCE

A brilliant anthropologist who used feminine intuitive powers to explore the Regional Underworld was the late Dr. Felicitas Goodman. I found I couldn't stand male-dominant anthropology courses in the university, and then I met Felicitas. We need some radical anthropological revisions at this time, and woman's wisdom is essential in this regard. Goodman has opened up a refreshing new anthropological vista, and as you may have guessed, orthodox anthropology simply marginalizes her discoveries.

I studied with Goodman from 1994 until she died in 2004, when she was ninety-one years old. She discovered a series of ritual postures that, for at least the last forty thousand years, were used by people as a tool for staying attuned to nature. Presently I'm an instructor for her anthropology institute, the Cuyamungue Institute of New Mexico,[49] where I offer workshops on ecstatic trance, a shamanic technique used for entering altered states of consciousness. One of the reasons I like this technique is that it allows anyone from any culture to experience other worlds. And my work with ritual postures has given me the best insights about the Regional people.

Goodman discovered that some ancient figurines and rock paintings are "ritual instructions" for entering "the alternate reality," the term she uses for the other worlds.[50] When a person assumes a specific posture combined with rhythmic stimulation such as rattling or drumming, and goes into a trance, Goodman wrote, "the body temporarily undergoes dramatic neurophysiological changes, and visionary experiences arise that are specific to the particular posture in question."[51] During normal learning, the negative charge of the brain is 250 microvolts, yet when in trance some people's negative charge goes up to 1,000 to 2,000 microvolts.[52] I find the trance experience activates right-brained consciousness, and brain-wave testing also seems to suggest this, since electrical brain activity goes into the theta range.[53] Maybe this is because the left hemisphere of the brain balances with the right hemisphere when we are in the theta range.

At the Cuyamungue Institute, we call this experience "ecstatic trance," and truthfully, this practice is what has given me the courage to attempt to penetrate the consciousness of the Regional Underworld. I am in the middle of a research project that explores the Underworlds by means of ecstatic trance, and someday I may report on some of my

Fig. 3.12. *Painting in the ritual cave of Lascaux, France. The ithyphallic man demonstrates a ritual body posture from fifteen thousand years ago!*

findings. Whatever is going on in trance, it can be a direct access to the Regional world consciousness when we use postures from more than twelve thousand years ago. The ritual postures found by Goodman are mostly only a few thousand years old, yet some of these are from hunter-gatherer cultures, since either the cultures that invented them were in a hunter-gatherer phase, or because horticultural cultures inherited them from long ago. Also, I suspect that many of the postures from the last three thousand years are actually carefully guarded newer versions of old relics. We've found only a few that go directly back into the Paleolithic, such as the Venus of Galgenburg posture from thirty-two thousand years ago, the Venus of Lausel posture, which goes back twenty-five thousand years, and the Lascaux Cave posture, which goes back fifteen thousand years.[54]

Fig. 3.13. *The Venus of Galgenburg.*

Look at the illustration of the Venus of Galgenburg and imagine traveling with her back into her world. While assuming Paleolithic postures, the energy in a person's body is much more intense than anything I've ever felt. Many people can't even hold these really ancient postures for more than a few minutes. The Regional people must have been living with very intense levels of energy in their bodies and in nature that we don't feel these days. Certainly, the discoveries of Pierre Mereaux and Paul Devereux lead us in that direction.

You may be wondering where on Earth I'm going next! There is much more evidence for advanced consciousness during the global maritime civilization and back into the Regional Underworld, as well as for advanced cultures from 9000 to 3000 BC. It is hard for me not to present more of this databank, because it is just so fascinating. If you want to know more, you can read some of my previous books and those by the new-paradigm writers. As more of the ancient knowledge comes through by means of archaeology and anthropology, as well as by using ritual postures, I will probably keep reporting on it.

In the next chapter, we consider the influence of the 1998 galactic synchronization, which is encouraging all humans to evolve twenty times faster during the Galactic Underworld that began January 5, 1999. Since we will be evolving another twenty times faster during the Universal Underworld in 2011, it is time for all of us to come up to speed and learn how to vibrate with nature again.

4

ENTERING THE MILKY WAY GALAXY

THE 1998 ALIGNMENT OF THE WINTER SOLSTICE SUN WITH THE GALACTIC PLANE

At some time during 1998, the winter solstice Sun moved into alignment with the plane of the Milky Way Galaxy. As I explained in chapter 1 and will summarize here, John Major Jenkins hypothesizes that as far back as 100 BC, Maya astronomers at Izapa were calculating the location of the winter solstice Sun and its approach to the Galactic Center. After about one hundred years of observing the location of the winter solstice Sun on the ecliptic, these astronomers calculated that it would cross the galactic plane close to the Galactic Center in approximately two thousand years, or around 2011–2012. Jenkins posits that with the end of the Calendar as a target, they constructed the Long Count Calendar based on the beginning of civilization and its coming end.

Since the closest alignment of the solstice meridian with the galactic plane—the galactic-plane ecliptic crossing—actually occurred in 1998, the Izapa astronomers were only about fourteen years off, which is amazing. It turns out that if we factor in Calleman's theory of time acceleration, they were much smarter than we can imagine: The 1998 alignment actually created physical changes in Earth that may have been required for the next evolutionary acceleration times twenty during the Galactic Underworld! The beginning of the Galactic Underworld on January 5, 1999, is when our consciousness sped up twenty times faster than during the Planetary Underworld (AD 1755–2011). I will explore the possibility that the Maya knew it would take the alignment to the Galactic Center to shift humanity to this fast level of change that would come at the end of the Calendar.

As you will see in a moment, when the alignment occurred in 1998, there were monumental geophysical and astrophysical changes in Earth and the universe. Using 9500 BC as a dichotomy in the evolutionary timeline, some of the changes in 1998 may have shifted patterns in Earth that were set into place only 11,500 years ago. In other words, *the galactic alignment is having a huge effect on our planet's ongoing adjustment to the cataclysm,* especially with regard to our consciousness. Even beyond that, I will explore the possibility that this period from 9500 BC through AD 2011 is described in the sacred scriptures of India—the Vedas—and that the Maya and Vedic knowledge must be from the same source. These are radical ideas, yet the Galactic Underworld is a very radical time; it is almost impossible to go too fast with your mind.

DAY SEVEN OF THE NATIONAL UNDERWORLD

Day Seven of any Underworld is when we experience the fruition of that Underworld, the time when the essence of that whole acceleration becomes very apparent. The accompanying table shows the creations during Day Seven of the first seven Underworlds. Of course, we will not know what the Day Seven creations of the Galactic and Universal Underworlds are until 2011.

Considering the 5,125-year-long National Underworld, Day Seven began in AD 1617. In 1648, the Westphalian Peace Treaty was signed; it ended the Thirty Years War, which had begun at the opening of Day Seven. This war and eventual treaty birthed the modern nation-state by recognizing the *sovereignty principle,* the recognition of the rights of individual nations, which was the fruition of the 5,125-year-long evolution of historical civilization. Concurrently, modern science awakened in the seventeenth century. Just before we entered Day Seven of the National Underworld, Giordano Bruno was burned at the stake in AD 1600. Bruno was a member of the group of new scientists who were certain that Earth orbits around the Sun—heliocentric theory. And in 1619, Johannes Kepler published *De Harmonie mundi,* which Calleman notes was the first time higher mathematics was used to formulate nature's laws.[1]

Cycle	Beginning of seventh day	Highest Expression (scientific date)
Universal	AD 2011	?
Galactic	AD 2011	?
Planetary	AD 1992	Computer networks (1992)
National	AD 1617	Modern nation (1648)
Regional	8000 YA	Agriculture (8000 YA)
Tribal	160,000 YA	*Homo sapiens* (150,000 YA)
Familial	3.2 MYA	*Australopithecus afr.* (3.0 MYA)
Mammalian	63.4 MYA	Placental mammals (65 MYA)
Cellular	1.26 BYA	Eukaryotic cells (1.5 BYA)
(YA = YEARS AGO, MYA = MILLION YEARS AGO, BYA = BILLION YEARS AGO)		

Fig. 4.1. *Day Seven of the Nine Underworlds. (From Calleman,* Solving the Greatest Mystery of Our Time: The Mayan Calendar.*)*

This new but actually very ancient perspective meant that people had to realize that their current understanding of the heavens was a total illusion; the Sun was the center, not Earth. Also, this realization was the beginning of remembering much more ancient knowledge, since there is plenty of evidence that the Classic Maya, the Egyptians, Vedic civilization, and other ancient cultures knew that Earth orbits the Sun. These cultures even knew that our solar system travels around the center of the Galaxy, and they believed that our location in relation to the Galaxy influences events on Earth, as you will see in this chapter.

During Day Seven of the National Underworld, Europeans awoke from the Dark Ages, which was a time of massive regression for many people in the West. Let's imagine what life was like for them. After the Roman Empire fell apart during the fifth century AD and the West regressed, imagine how secure and grounding it was during the Dark Ages to believe that you lived in the very center of God's universe.

As Europe woke up during the seventeenth century, it began to dawn on a few scientists that Earth was spinning around the Sun, which was moving through space in an immensely expanded universe. This idea was very threatening, so the Vatican used Bruno as a scapegoat, making it easy for the other scientists to imagine being roasted. This meant that when

Copernicus, Kepler, Galileo, and others worked on their calculations for the new astronomy—celestial mechanics—they were totally alone in their new quest. They were very excited as they studied globes and made calculations, but they had to communicate with each other in secret.

Because of what happened to Bruno, the theories of these visionaries were potentially heretical and thus rejected out of hand by the orthodox intellectual and cultural authorities of the time. Now we know that these radical theories were correct. And while humans did gradually adjust to the idea that Earth was not the center of the heavens, ironically all they were really doing was recovering an ancient science from many thousands of years ago!

GALACTIC CONSCIOUSNESS SINCE 1998

Physicists, astronomers, and astrophysicists have been successfully mapping the universe since the 1950s. Much of the public knows that our Sun revolves around the center of the Milky Way Galaxy in a 225-million-year orbit, and that it is approximately thirty light-years out from the center. Our Galaxy is part of a cluster of galaxies called the Virga Supercluster, and there are many more superclusters in the universe.[2]

Mysteriously, *cosmologists discovered in 1998 that the expansion of the universe was suddenly speeding up!* This shift may ultimately provide the proof for superstring theory (the idea that the universe is made of vibrating strings functioning in ten dimensions) because it suggests that *quintessence*—a strange energy that can exert an antigravity force—somehow "switched" in 1998. That is, cosmologists considered the possibility that *quintessence started accelerating the cosmic expansion in 1998!*[3] This has to be an effect from another dimension, which in my model would be the ninth dimension, the dimension of the Mayan Calendar.

Cosmologists have penetrated the real nature of our Galaxy and our place in it just in time, because re-attaining galactic consciousness is the agenda of the Galactic Underworld. It looks as if some fundamental cosmological forces changed in 1998, which helps me understand how the acceleration times twenty, which began January 5, 1999, was even possible. And what changes will there be in the universe when the Universal Underworld acceleration kicks in in 2011? Science is the language of our times, so the more information forthcoming about the

Milky Way, the better. As you will see in this chapter, it is highly probable that the Maya and other highly developed ancient civilizations may have known more about the Galaxy than we do right now. Regarding evolutionary acceleration times twenty, now it is time to comprehend the time distortions of the Galactic Underworld stacked on top of the Planetary Underworld (AD 1755–2011), stacked on top of the National Underworld (3115 BC–AD 2011), and so on. The Planetary Underworld accelerated our climb out of the slower time of the National Underworld by speeding up time by twenty times; this recovery of an accurate scientific view of the universe has become ubiquitous since 1992 by means of computer networks. Meanwhile, during Day Seven of the National

Fig. 4.2. *The nine-story cosmic pyramid symbolic of the Nine Underworlds of creation. Each of these develop a certain level of consciousness, and all reach their completion on the day 13 Ahau, October 28, 2011. (Adapted from Calleman,* The Mayan Calendar and the Transformation of Consciousness.*)*

Underworld and all of the Planetary, *materialistic science has eliminated the spiritual understanding of science.*

This approach will not hold up during the Galactic Underworld because both the material and spiritual approaches to science are needed now. Regarding Calleman's Planetary Round of Light mentioned in chapter 3, the Galactic Underworld is a dualized phase that favors right-brain access, which is very spiritual.[4] This is why an interest in spiritual science is increasing in our culture. An example of this is the wild popularity of Dr. Masaru Emoto's *The Hidden Messages in Water.*[5]

As I previously noted, I find it fascinating that the Coba stele was deciphered in the 1950s, when the scientific evolutionary timeline was also assembled. The flowering phase of the Planetary Underworld in Day Six just happens to have been from AD 1952 to 1972: the 1950s again! We can understand more about the convergence of so many scientific theories at this time by having a closer look at the Planetary Underworld, which will help us comprehend the nature of the speedup times twenty within the National Underworld.

THE PLANETARY UNDERWORLD ACCELERATION: AD 1755–2011

Calleman notes that the best way to see the influence of the speedup by twenty times in 1755—right in the middle of Day Seven of the National Underworld—is to follow the development of telecommunications, which he does in figure 4.3.

Just like the 820-million-year Mammalian Underworld, in the Planetary Underworld creation comes forth during the Days, and integration occurs during the Nights. As you can see, the telegraph appeared in 1753, and by the time we evolved to 1952 at the opening of Day Six, we had the first public television broadcasts. I will never forget that moment. We had a grand piano in our living room. I had begun lessons when I was four or five years old, and by the time I was nine in 1952, I was studying with a great teacher. But never mind all that, when television came on the scene, out went our piano and in came the TV!

Is there anything that has changed our basic way of life more than television? And think about how computers have changed our lives. As Calleman notes, the Planetary Underworld was when industrialization was seeded, and communication by written message during the National Underworld was just not fast enough anymore.[6]

Because of the speedup, there was a push to develop technology, and a common communications network—the Internet—was needed. This push that began with the telegraph has evolved all the way to the development of the Internet; it drives this new level of human creativity. It is my belief that quantum technologies will be in place by 2011. Figure 4.3 really verifies the acceleration during the Days of the Planetary Underworld, just as we noticed with the Days of the Mammalian Underworld; the concordances are startling and meaningful.

Growth Stage Day and Heaven No. Ruling Deity	Time Span	Invention or Development
Sowing Day 1, Heaven 1 god of fire and time	1755–1775	**Theory of telegraph** Anonymous (1753) Bozolus (1767)
Germination Day 2, Heaven 3 goddess of water	1794–1814	**Optical telegraph** Chappe, Paris-Lille (1794) Sweden (1794)
Sprouting Day 3, Heaven 5 goddess of love and childbirth	1834–1854	**Electrical telegraph** Morse (1835) Washington-Baltimore line (1843)
Proliferation Day 4, Heaven 7 god of maize and sustenance	1873–1893	**Telephone** Bell's patent application (1876) First telephone station (in U.S., 1878)
Budding Day 5, Heaven 9 god of light	1913–1932	**Radio** First regular broadcast (in U.S., 1910; in Germany, 1913)
Flowering Day 6, Heaven 11 goddess of birth	1952–1972	**Television** First public broadcast (in U.K., 1936) First color TV broadcast (in U.S., 1954)
Fruition Day 7, Heaven 13 Dual-Creator God	1992–2011	**Computer networks** Internet (1992) Global television channels Mobile telephones

Fig. 4.3. *The evolution of telecommunications during the Planetary Underworld. (Illustration from Calleman,* The Mayan Calendar and the Transformation of Consciousness.*)*

Seeing that the 1950s and 1960s encompass Day Six of the Planetary Underworld (the flowering of communication systems), we can see why so many complex databanks were being formulated during the 1950s. This outburst of creativity required strong links between all people—television got the news out to the planet. Remember the good old days when you could get important information on television? Now, only the Internet and cell phones are fast and free enough for people. Yet, just as television became controlled during the 1990s, global elite controls are being put on the Internet and cell phones as I write these words.

However, unlike television, now the time acceleration is too rapid to stop the flow of information, and next quantum communications will take over, since people are becoming more telepathic. This is because of the dualistic and right-brain nature of the Galactic Underworld, and because of what happens to human consciousness when people are galactocentric. Just like the incredible brilliance of the Mayan Calendar, humanity is right in the middle of an overwhelming intellectual waking-up process. I prophesy that in 2011 during the Universal Underworld we will all be psychic, just as people were thousands of years ago during the Regional Underworld.

To really understand time acceleration, the speedup by twenty times in 1755 during Day Seven of the National Underworld is the best access we have to feeling how time acceleration works. This is because many of us know a lot about the Planetary and National Underworlds, and most people have been feeling very peculiar since 1999. Think of how strong and effective the change is that began as recently as 1755 with industrialization, and how much faster 1755 (and beyond) seems to be than AD 1500. Then if you dare, look back to 1999 when the world became a strange place once the energy of the Galactic Underworld came on strong.

In January 1999 in the Kripalu Chapel in Lenox, Massachusetts, my students and I conducted the first Pleiadian Agenda Activation with newly composed music by Michael Stearns. After that we all enjoyed a concert in the chapel with a group called Galactic Gamelan from the Massachusetts Institute of Technology, which had been trained by a Balinese gamelan master. None of us knew they were appearing right after our activation, so I extended our class to include the concert. I was shocked to see that behind the altar of the chapel, which was crammed with statues of Indian saints, was a very tall mosaic composition of St.

Ignatius Loyola standing on a high bluff and pointing at the Pleiades! This event was magical, and we all felt something entirely new was going on. These kinds of reflections back to early 1999 offer much assurance that the convergences happening in science and in Mayan Calendar research are right on schedule. There is a virtual explosion of creation going on now that is driven by the intelligent designer, and I find Calleman's time-acceleration factor to be the only explanation that helps with the dizzying pace of things. It is a good idea to go back to January 5, 1999, and reflect on what you have been doing from that point on, based on the Mayan 360-day tun system. I have provided a pyramid for just that process in appendix C with instructions on how to use it.

GALACTOCENTRIC ASTRONOMY AND ASTROPHYSICS

Because this chapter is devoted to understanding as much as possible about humanity becoming *galactocentric* during the Galactic Underworld, we will return to a contemplation of astronomy and astrophysics. While this new cosmology is even more radical for us to accept than the notion of heliocentricity, the ancients knew that *penetrating the Galactic Center is the real secret to entering the universe.*

As you probably know, when you look into deep space, you are looking way back in time. To date, astronomers with modern telescopes have looked back 15 billion years, and in 2011 they will look back to 16.4 billion years. That is, the farther astronomers reach out into space with their telescopes, the farther back in time they go. When I've tried to understand what it really means to say you can look back to the beginning of creation, I've been stumped by the concept, although I understand it is based on the speed of light. Maya shamans said they were looking back to the creation events, and then they elucidated the timing of our unfolding until now. Specifically, Maya shamans said we are unfolding with the divine mind through various cosmic eras in an evolutionary process of cyclical births, destructions, and rebirths. As Douglas Gillette puts it in his fascinating book *The Shaman's Secret,* "The universe of our most advanced sciences is disturbingly similar to the cosmos the Maya imagined."[7]

I began this chapter attempting to show a little bit about the literally fantastic changes the human mind has been undergoing during the last four hundred years. I offer these ideas because the degree to which we need to integrate our thought during the Galactic Underworld is truly mind-boggling. The creation of modern communications began in the Planetary Underworld, and when the television created a global mind-device, suddenly the public was asked to visualize Earth in a new place in the universe, pummeling through the darkness of space—yet we were all just monkeys a little while ago!

Since the 1950s, we have all been deeply contemplating our identities and places in the universe, and we are in the middle of an incredible breakthrough that is so mystical that people will soon drop their differences. For example, the contemporary world looks very different when you factor in the nature of the Galactic Underworld. Its right-brain influence is causing an Eastern spiritual awakening, while the West loses its dominance.

Because the Galactic Underworld is dualized, initially this process has been characterized by extreme Western aggression against the more spiritual East. For example, by invading a sovereign nation (Iraq) in 2003, the Bush administration dumped the sovereignty principle gained by the Westphalian Peace Treaty of 1648. However, increasingly through the latter stages of the Galactic Underworld, Western dominance will break down and East and West will balance. Similar to what happened during the Thirty Years War, the aggressors will run out of steam. I mention this regression here because I would like you to take a moment to feel how much you actually object to the scurrilous political scene while your mind is expanding out into the Galaxy. So, let's go!

THE 1998 GALACTIC-PLANE ECLIPTIC CROSSING

The really big news in the world is that as of 1998, *the winter solstice Sun moved into alignment with the plane of our Galaxy!* This special energy so generated, with its new winter light, is helping us discover our place in the Galaxy, and we are experiencing maximal levels of energy from the Galactic Center during this experience. Lest anybody think this is New Age "woo-woo," astronomers call this the alignment of the solstice meridian with the galactic equator, which is shown in figure 4.4.

Because we are talking about the plane of the Galaxy—the galactic equator—there is some disagreement among astronomers about the

Fig. 4.4. *Alignment of the solstice meridian with the galactic equator. (Illustration adapted from Jenkins'* Galactic Alignment.*)*

exact date that this alignment occurred; however, most agree that it was the middle of 1998. The U.S. Naval Observatory said the exact alignment occurred October 27, 1998, and English astronomers concurred that it was in 1998, but some months earlier, on May 10; they even had a big party on that day to celebrate it.[8] I've already written extensively about this alignment in *The Pleiadian Agenda, Catastrophobia,* and *The Alchemy of Nine Dimensions*. John Major Jenkins has also written much about it, as have other researchers.

What matters here is that the astrophysical alignment was in 1998, which was verified by some actual changes in Earth. In *Alchemy,* I noted that this alignment is occurring while our solar system is at *perigalacticon* (when our solar system is the closest to the Galactic Center).[9] That is, we are experiencing the most intense contact with the powers of the Galactic Center in modern times; 1998 was the apex of this intensity, and this is a huge confirmation for Calleman's date for the opening of the Galactic Underworld right at the beginning of 1999. Of course, this means that in 1999 time sped up by twenty times over the Planetary

Underworld, which was detectable; the changes in Earth are the big news about the galactic alignment.

EARTH DEVELOPS A BULGE IN 1998

During 1998, there was much evidence for a huge shift in Earth during the alignment. From 9500 BC until AD 1998, high-latitude regions of the Earth have been rebounding from the weight of the glaciers (or tectonic shaping effects of the cataclysm), which caused Earth's mass to shift gradually to the poles. Suddenly in 1998, the gravity field began getting stronger at the equator and weaker at the poles, and Earth's rotation slowed slightly.[10] Then a mysterious bulge formed at the equator![11] This is an absolutely unprecedented change for our planet since 11,500 years ago.

Back in 1995, when I channeled *The Pleiadian Agenda,* the Pleiadians spoke of many radical changes during 1998. They said, through me, that in 1998 "samadhi waves would begin to alter nature radically."[12] Imagine my surprise when in August 1998, a tremendous X-ray and gamma-ray bombardment from a collapsed star blasted Earth. For ten minutes the sky was writhing with light, which shut down a lot of scientific instruments. Most significantly, astronomers said this moment was the "first observed physical change to our atmosphere from a star other than our Sun."[13] Before I knew anything about the Galactic Underworld, I said in *Alchemy of Nine Dimensions,* "We humans experienced a jolt of high energy in 1998 that may end up being viewed as the *initiator of a new stage of evolutionary consciousness.*"[14] In a moment, we will see what that new stage of evolution could be.

The period from January 5, 1999, through October 28, 2011, is of such significance that chapters 6 and 7 are devoted to the Galactic Underworld. Before that, here I am placing the Galactic within the context of the larger whole—the new alignment of our solar system in the Milky Way Galaxy. The 1998 alignment seems to have created a new frequency field on our planet, probably a magnetic field, which is making it possible for us to withstand and integrate the changes of the Galactic Underworld. For example, regarding the bombardment in August 1998, scientists noted that similar X-ray cosmic blasts close to Earth might have been the cause of Earth's extinctions in the past, which also often correlate with fast stages of new evolution, such as the Cambrian period 570 million years ago (567.9 million years ago according to Calleman).[15]

I surmise that *there may have been an extinction of some forms of human consciousness in August 1998,* the forms of consciousness that cannot survive in such an activated environment. Exactly what this extinction could be will only be apparent in the future. However, as things go, the more we understand the nature of these changes, the easier it will be to go with the flow instead of resisting it. Our species is being influenced by astronomical forces that have never occurred on this planet before, and certainly not while humans have been around. The amount of intellectual integration by humanity during the last fifty years is incomprehensible for most people, and defensive fundamentalism is a reaction to all this change. I understand that and sympathize with the problem, but that does not diminish how retrograde and dangerous fundamentalism is. Of course, this includes Islamic fundamentalism.

Anyone in America with a nonthinking sister, brother, parent, or child knows that the rise of fundamentalism is causing great pain. As a result of writing so much about this issue and having the opportunity to observe how our students are handling the stupendous galactic synchronization, I want to spend some time discussing an entity that we have to befriend now—the black hole in the center of the Milky Way Galaxy. The black hole is consummate darkness, and we must befriend the darkness lest it consume all that is on Earth. Fundamentalists are afraid of darkness, so they are creating huge shadows that threaten all life because *their big shadow is adoration of war.* The very basis of fundamentalism is that if a person is right, killing is justified. This belief is not the truth, it is retrograde from an evolutionary perspective, and as you can see in figure 4.5, that form of human adaptation is ending now.

BLACK HOLES AND SINGULARITIES

A black hole forms when a large star experiences a rapid gravitational collapse and gets sucked into a funnel, which is shaped by the curvature of spacetime. Once matter is sucked into this funnel, it is crushed into unimaginable density. Yet, since matter is energy, it becomes a singularity—a point of zero size—and then this matter appears in a "different universe," which can be thought of as a new universe, something that the late Itzhak Bentov, who was a brilliant explorer of physics and

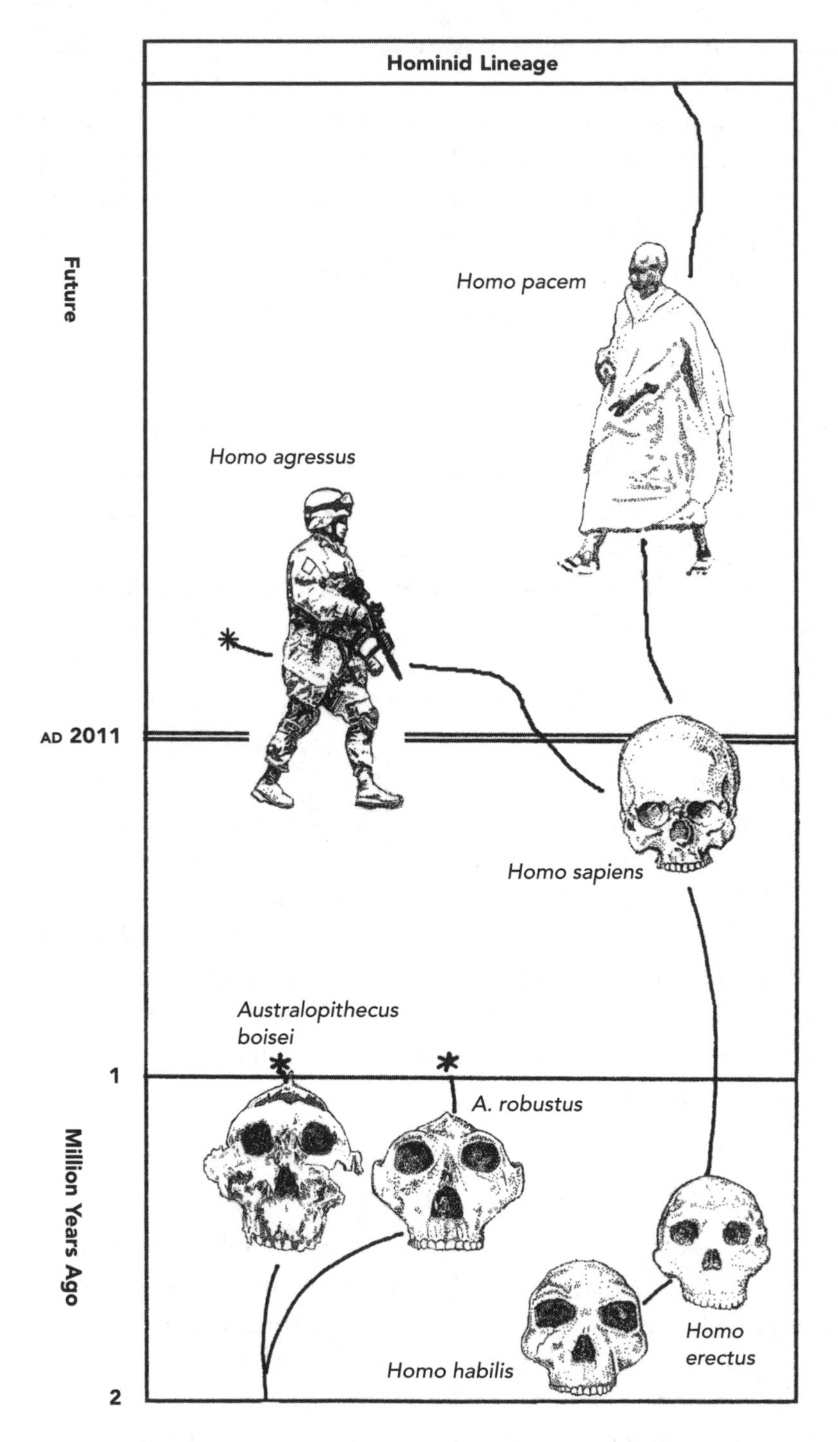

Fig. 4.5. Homo pacem. *The asterisk indicates evolutionary strains going out of existence (see fig. 2.9).*

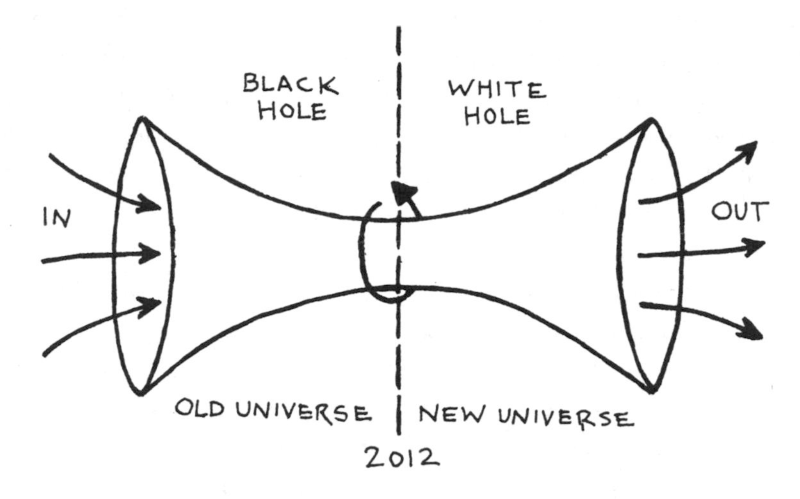

Fig. 4.6. *The new universe. (Illustration adapted from Bentov,* Stalking the Wild Pendulum.*)*

consciousness, thought of as a "white hole."[16] Figure 4.6 shows this concept, with 2012 penciled in at the singularity point, which is one way of imagining where we may soon be headed.

Figure 4.6 has 2012 instead of 2011 for the singularity because 2012 is when Earth's time acceleration is complete, and during 2012 the acceleration will be in the Milky Way Galaxy. Bentov notes that as we move in space, we are also moving on a time axis, which is exactly what the Mayan Calendar time wave is. What is expanding is our spacetime (which I think is akin to the 20 × 20 acceleration factor), and the greatest rate of expansion occurs at the point where matter reverses direction.[17]

The outside edge of the black hole is called the "event horizon," which is a boundary beyond which light cannot escape. Once matter crosses it, it is sucked down the hole. If you could look back after you had been sucked in, you would see the future history of the universe flash before your eyes; yet once inside the black hole, you'd be unable to communicate anything you saw back to anyone outside it.[18]

Considering this wild description of a black hole, scientists have to look at cosmology this way by the laws of modern physics, yet they admit their own description befuddles them. What really matters is how each one of us might feel in a black hole, since each one of us is an individual. As you approach the singularity (where you come out in another universe), you'd feel yourself being torn apart atom by atom. At the singularity, everything you've ever known about the universe breaks down. When you pass through, you wake up as though you've been reborn as an adult—and there you are in a new world, the white hole. The illustration may help you imagine what happens to somebody who goes too fast while flipping on his or her skateboard and accidentally falls into a black hole.

I love watching the kids with baggy pants hanging off their butts as they flip up and around on skateboards or snowboards. I think they are teaching themselves about black holes! To me as an adult, the Galactic Underworld feels like the early stages of falling into a black hole, and I find myself feeling like I am skateboarding. Problem is, if I got on a skateboard with my white hair, I'd be arrested by the nearest cop. How about you?

Notice that this process we have just described involves time distortion. I think Calleman's model captures the essence of time distortion because all nine levels evolve simultaneously, and they are stacked upon each other. The animals 820 million years ago probably had trouble handling the speedup of the Cellular Underworld! Yet, the minute one factors in nine dimensions of consciousness, as with my nine-dimensional

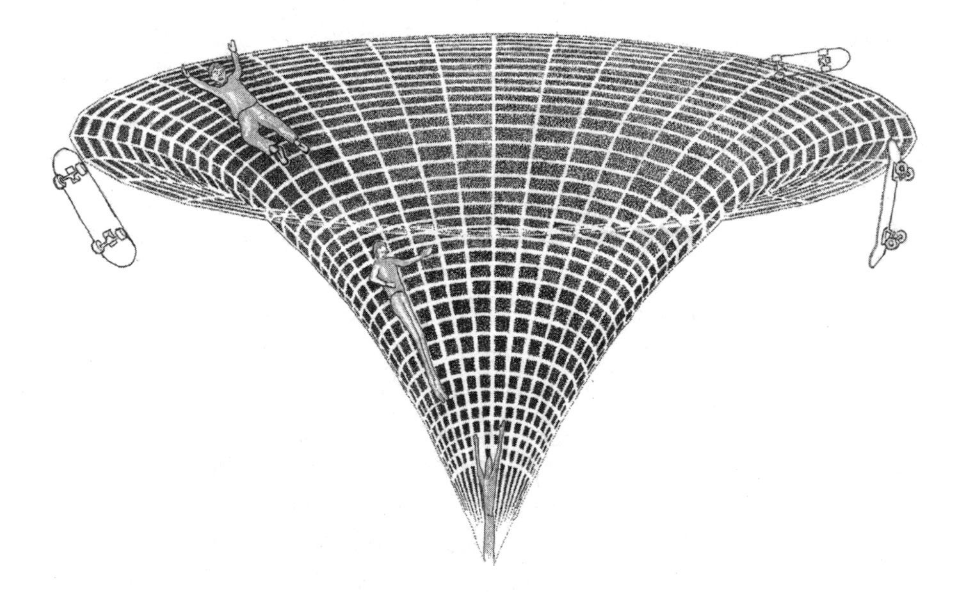

Fig. 4.7. *Falling into a black hole.*

model, it becomes possible to imagine this, which we will do later. In the meantime, we still have more to learn about black holes.

In *The Pleiadian Agenda,* the Pleiadians said the black hole in the center of our Galaxy is the source of time on the nine-dimensional vertical axis, the line that connects all nine dimensions. They said the black hole is a *spinning gravitational nucleus that manifests itself in time waves, which create events on Earth.*[19] They said that once the black hole started really activating us (in 1999), we would be forced to process deep trauma and go beyond just a survival mentality, which is exactly why the Galactic Underworld is so intense. Another way of putting this would be to say that something is going on during the Galactic Underworld that makes it possible for humans to feel the time dynamics of black holes.

In 1998, physicists suggested that gravity is capable of being one of the four fundamental forces because it is diluted by its propagation through the unseen dimensions. This suggests that gravity may be the force linking the unseen dimensions that may cause time acceleration, and that gravity is the link to these unseen dimensions.[20] Think of the black hole as a spinning gravitational nucleus creating time waves that propagate events throughout the unseen dimensions. If you can get there at all, I'll tell you where my mind goes with that idea: It goes right to the thought of a Maya shaman two thousand years ago who traveled into the black hole and brought back his vision of the future—the Mayan Calendar.

Science says once you're sucked into a black hole, you see the future history of the universe, but that you wouldn't be able to communicate it. This is where science really falls down by being materialistic, since *the only way you can go into a black hole is in your consciousness.* Consciousness is not limited by time or space, so in your mind you *can* travel anywhere in the universe, *and you can return.* This fact is the only thing that explains the brilliance of the Maya. As you will see in chapter 8, I have not yet scratched the surface of what the Maya had figured out. What needs to be absorbed here is that *accessing the intelligence of the Milky Way black hole is the essence of the current stage of evolution.*

Once he saw the future, our Maya shaman was still able to bring it back, and he was able to communicate the divine plan of the universe that functions by time acceleration. I think the concept of nine accelerating Underworlds predicts the future through 2011, and we are playing out the cycles that are spinning out of the Galactic Center. This must be why back in 1995 the Pleiadians put time in the ninth dimension and called it Tzolkin (the Day Calendar), the highest dimension humans can access at this time. That is, the great Creator is time, which is in the highest dimension according to the Pleiadians. Since we have the capacity to communicate with any intelligence in the universe, we can communicate with this being, the Creator. The Maya have proven this, and we are just figuring it out now. For example, the Ninth Underworld, which may correspond with the ninth-dimensional Tzolkin, is only 260 days long, which is just one round of the Tzolkin.

Materialistic science denies the possibility of accessing intelligence this way, yet the level of evolutionary accuracy of the Mayan Calendar (as interpreted by Calleman) proves this *is* possible. How else could the Maya have known the dates of the geological and biological evolutionary timeline or the creation of the universe? How else could they have realized what would happen to Earth after 1998? During our activations, our students travel into the Galactic Center and the black hole all the time to gather information. Yet, many of them are still having a hard time doing this these days because of the sheer volume of emotional processing being caused by the speedup.

Why is it so hard to handle this speedup? We are beings who function with four bodies of consciousness—the physical, emotional, mental, and spiritual bodies. Because of the cataclysm in 9500 BC and the follow-up adjustments, our emotional bodies are filled with unprocessed fear and trauma, the theme of *Catastrophobia.* I have become convinced that Earth had a vertical axis as it orbited the Sun before the disaster. Allan and Delair, as well as some others, believe that our axis got knocked into a 23½-degree tilt only 11,500 years ago. In other words, *we not only became an icosahedral sphere with adjusting tectonic plates, we also became a tilting planet.*

THE TILTING ICOSAHEDRAL EARTH

Because Allan and Delair's evidence for such a recent axial tilt is so convincing, I adopted their idea as a working hypothesis while writing *Catastrophobia,* yet I was by no means able to prove it. I still can't; it would take a large group of astrophysicists to do so, since the whole solar system is involved. There are many mythological sources and records from indigenous people suggesting that within recent memory Earth's axis was vertical. Appendix A, which covers the scientific evidence for recent axial tilt, comes from *Catastrophobia.* I've included it because I think the information is of monumental importance, and because I need to mention a few things about it here.

Notice in appendix A that Allan and Delair said that "an equatorial bulge would remain as an essential stabilizing feature" if Earth's axis did *not* tilt (that is, if it were vertical). As already discussed, Earth developed an equatorial bulge in 1998 when the Galactic alignment occurred. I've often wondered whether our planet is slowly gyrating back to a vertical-axis position, and the 1998 equatorial bulge supports this possibility. Is this what our planet would do if it is true that we went into a tilt only 11,500 years ago? Is that what the end of the Mayan Calendar is about? As many readers may know, the Earth's magnetic field has been dropping steadily for at least two thousand years, which leads many scientists to believe a polar flip may be imminent. Perhaps an axial alignment might be more likely?

In general, regarding appendix A, one simple thing that can be said about a vertical or tilting axis is that a vertical axis would be much more expected for our planet than a tilting one. Also, the existing tilt is what one would expect from a disaster like the one Allan and Delair describe. Assuming the axis was approximately vertical until 11,500 years ago, humanity has been adjusting to a radically different Earth since then. For example, if there was no tilt until the cataclysm, then *there was no precession of the equinoxes until 9500 BC*. Unfortunately, this is a complex issue for this book. But it must be included, because the very reason we are experiencing the galactic alignment is that precession is causing the winter solstice Sun to closely align with the Galactic Center. The point is, I think *precession is a very recent phenomenon for humanity.*

THE NEW PRECESSION AND THE RIG VEDA

If it is true that precession began only 11,500 years ago, then the Sun's current winter solstice approach to the Galactic Center is the apex of this alignment. This alignment is what is projecting human consciousness into the Galaxy. I think humanity began a new relationship with the Galaxy only 11,500 years ago, which leads to all kinds of ideas. For example, this would mean that the planet is very different for us now than it was during the Paleolithic global maritime civilization. There is every reason to believe this is the case, and I think this is the reason the Elder Culture instructed the people about the newly tilting planet. The essence of the Elder Culture can be found in the Vedic records because the people of India have been so faithful in protecting their cosmic knowledge, which is why we have direct access to that knowledge today. *Veda* means "knowledge," while *maya* is the Vedic word for "the illusion of time," the Nine Underworlds!

There is good evidence that an Elder Vedic Culture goes back as far as 11,500 years ago. The Indian author B. G. Sidharth decoded the astronomy of the Vedas, the world's oldest sacred literature.[21] He was able to date the Rig Veda, the oldest of the Vedas, back to approximately 10,000 BC. This date is so close to 9500 BC, that I checked the author's method of dating. He notes that this date is "12,000 divine years" or the Great Age or the Mahayuga.[22] Vedic divine years are 360 days long instead of 365 days long, the same as the Maya divine year. I hope that got somebody's attention out there. In other words, *the Vedas are based on a tun-based Maya divine year!* So, I calculated twelve thousand years of 360 divine years (instead of 365 solar years), and that solar date is exactly 9400 BC! Therefore, this means the Vedic Mahayuga began right after the cataclysm, when I am saying precession actually began. This means that, among other subjects, *the Rig Veda may have been written to describe precession,* a totally new phenomenon. The effects of precession in the sky (as well as the tilting axis) would have been quite startling to the people who survived the disaster. Describing it was necessary to help people understand the new astronomy, which may be the inspiration for the earliest surviving fragments of the Rig Veda. Sidharth concludes that the Rig Veda was "composed by a highly intelligent people with a fairly advanced working knowledge of astronomy, which was enlightened enough to include concepts like precession, heliocentrism, and the sphericity of Earth, and which matches the astronomical knowledge of the seventeenth century CE."[23] Sidharth is a highly respected Indian scholar, and his conclusions are very well proven.

The Vedas are of great interest here because both the Maya and Vedic cultures worked with vast dates that go back millions and billions of years, and because both cultures knew about precession and factored it in. Yet Vedic astronomers used precession to analyze the rise and fall of cultural cycles much more than the Maya did.

A fairly recent precessional analysis of the Vedic scriptures by Swami Sri Yukteswar, a beloved and knowledgeable saint in India who died in 1936, posits that the rise and fall of human societies are determined by where the stars are circling around the poles during twenty-six-thousand-year-long precessional cycles.[24] Applying Calleman's theory, the Maya had a completely different sense of time wherein time accelerates by factors of twenty during Nine Underworlds that come to an apex in 2011. In other words, *Vedic cycles are repetitive, while Maya time accelerates to an end phase.*

Some Vedic long dates are close to Maya long dates, such as the fact that 4.32 billion years is close to being one fourth of the 16.4-billion-year creation date. It is possible that time acceleration by twenty once existed in the Rig Veda, which is a worthy idea for someone to explore. After all, Calleman discovered the time-acceleration factor in the Mayan Calendar only a few years ago. Sidharth notes that during the course of history, by the early centuries of the Christian era, Hindu writers had lost the memory of the previously enlightened Vedic astronomers, so such possible fragments may be gone.[25] If the Rig Veda was actually composed in 9400 BC, then it is of great import for this book, because it was probably one of the first tools for instructing traumatized humans after the cataclysm. Things like this inspire me to imagine just how advanced the Vedic and Elder cultures were, since modern astronomy only caught up to the astronomical knowledge in the Rig Veda fifty years ago!

If exponential evolution by twenty times does exist in the Rig Veda, then we'd have to be open to the possibility that the Maya got this idea from the Vedic civilization. This would have meant that the Vedic Elders gave the divine evolutionary plan to humanity right after the cataclysm, and that the Vedas influenced later Maya thought. In *Mayan Genesis*, Graeme R. Kearsley explored the voluminous evidence for Vedic cultural influence on the Olmec and Maya cultures.[26] I am suggesting that the purpose of the Vedic revelation could have been to communicate about the great changes in the sky, since Earth's axial tilting was suddenly causing seasonality, a totally new phenomenon. Also, the Rig Veda has many stories that sound like fresh memories of the cataclysm, such as legends about the churning of the oceans by gods and demons.

GALACTIC SUPERWAVES AND PRECESSION

If precession began only 11,500 years ago, then new dynamics exist that are connecting Earth with the forces of the Galactic Center. For example, the Galactic Center is about 26,000 light-years from Earth, and one full precession cycle is about 26,000 years long. This means that *the number of years it takes light to travel to Earth from the Galactic Center is about equal to one full cycle of precession.*[27] Is this striking fact just some kind of coincidence? I don't think so. Now that we are aligning with the Galaxy during the end of the Calendar, maybe the potential of this light-year synchronicity is activating Earth. John Major Jenkins has some very interesting ideas about this relationship. He notes that the astrophysicist Paul LaViolette says that this relationship (26,000 years/light-years) may be creating some kind of "entrainment between superwave bursts and the precessional cycle."[28]

Superwave bursts are LaViolette's hypothesis that great waves are periodically sent out of the Galactic Center. Jenkins says that LaViolette suggests that "when the earth's North Pole is tilted away from the Galactic Center, a superwave burst arriving at that time would propel the earth to wobble on its axis," and that these bursts may be entraining Earth's wobble to the superwave periodicity.[29] Jenkins notes, "The sense is that something emanating from the Galactic Center is responsible for precession."[30]

LaViolette has presented solid evidence for a superwave burst from the Galactic Center 14,200 years ago that he thinks may have triggered the Vela supernova (which, according to Allen and Delair, caused the cataclysm 11,500 years ago).[31] The research of Paul LaViolette is very germane to the discussion in this book because the Maya were definitely very concerned about solar and galactic influence on Earth during the end of the Calendar. Examining these ideas in detail at this point is too complicated, but I cover them in depth in chapter 8. What matters here is that the field that Earth functions within is significantly more intense since 1998, and the ancients somehow knew that this short period of time would be extremely significant. They went to a lot of trouble to inform their descendants—us—about these issues. Wanting to further penetrate these relationships, Jenkins goes into a fascinating discussion about the work of a brilliant twentieth-century philosopher, Oliver Reiser, who was captivated by the Galactic Center. Reiser believed that *human evolution is tied to the movements of the Galaxy,* and he was interested in the relationship between geomagnetic field forces and human evolution. According to Reiser, the field of energy above Earth that influences human evolution—what he calls the Psi-Bank—is where we receive Galactic information.[32] Reiser shows how biological evolution is driven by the changing field dynamics of the Psi-Bank, which are especially affected by cosmic ray showers that originate from the Galaxy. Regarding the previously described August 1998 gamma ray bombardment, the Psi-Bank must have changed in some way to accommodate the next proposed stage of human evolution—*Homo pacem,* the peaceful human, as already shown in figure 4.5.

In the next chapter, we explore the mysterious World Tree, the mechanism that, according to the Maya, is driving evolutionary acceleration. We investigate how the World Tree affects us as a species; we need to feel the World Tree in our bodies.

5

THE WORLD TREE

SACRED CULTURES AND THE WORLDS

Now that you have contemplated Carl Johan Calleman's time-acceleration theory that appears to be what the Mayan Calendar dates are describing—the activating basis for evolution in the universe—it is time to ask: What is the mechanism that drives this remarkable evolutionary system? According to Calleman, the driver of evolution through time is the World Tree, a magical tree described in the Popol Vuh. It is also the tree that shamans use to access all worlds.

In the Maya spiritual landscape, the World Tree generates the four sacred directions moving out from the sacred center—Yaxkin—a system for humans that shapes and accesses the spiritual worlds. In Maya land, the sacred tree is the ceiba, a very tall and beautiful tree that rises high to form great canopies in the jungle.[1] The analog in Celtic culture is the oak tree, and in Indian spirituality the banyan tree, under which the Buddha experienced enlightenment. According to mythology, the World Tree was created first in the universe, and everything emanates from it. In ceremonies, the Maya nurture the World Tree, which I have experienced many times.

Since 1995, I've been creating Pleiadian Agenda Activations, but it wasn't until 2005 that I realized that these activations resonate with what I call the "shivering" of the World Tree. When this insight came to me, I finally fully understood the purpose of my own work. Because all Nine Underworlds evolve simultaneously, the relationship is that the exponential speeding up of the World Tree frequencies causes the transformation of the worlds in the nine dimensions. Time acceleration is exceedingly challenging, and our activations seem to help students (as well as Gerry and me) integrate the rapid time acceleration of the Galactic Underworld—1999 through 2011. If Calleman is right about his interpretation of the Calendar, we humans are not just caught on a cyclical wheel of time that spins out repeating cycles. Instead, we are experiencing an ever-tightening spiral of evolution that is pulling us all into its own apotheosis—the transformation of the human into enlightenment. When I consider how much our students have evolved through this process, I am very optimistic about a positive outcome on planet Earth.

By necessity, this chapter is highly theoretical, since we are considering Calleman's theory of the central driving mechanism of growth and change on Earth—the World Tree. Let's start with the question: What is the general idea of the sacred tree, the World Tree? Sacred cultures have always used trees to organize Earth's intelligence, and shamans have always traveled in them to visit many worlds. What do I really mean by "sacred cultures"? Sacred cultures believe the material world emanates from the spiritual world, and they use key symbols to show how the spiritual world is organized. Sacred trees all have certain things in common—roots that reach down into the underworld, a great trunk in the middle world, and branches and leaves that reach the upper world, the cosmos. We can travel in them to access worlds because *sacred trees are the living structure of all the worlds*. Whether it is Yggdrasil in Scandinavia, the Sacred Tree of Life of the Kabbalah, or the Sacred Tree of the Celtic Middle World, all sacred science in ancient cultures thought of these trees as circulation systems for human consciousness. Eventually, all shamans learn to travel into the lower, middle, and upper worlds.

Regarding sacred postures and ecstatic trance—the most accessible shamanic tools I've found for modern people—we have specific techniques for traveling into the three worlds of the World Tree. I could write a book about the symbolism of sacred trees, but others already have, so here I will focus on a few aspects that are germane to the subject matter of this book. We will look into the World Tree of Maya creation mythology—the *axis mundi* of Earth—as the centering vortex that drives the systems of our planet. The sources are the Popol Vuh, the most important of the Maya sacred scriptures, and many Maya inscriptions and cosmological pottery that depict the Tree.[2]

THE MATERIAL AND SPIRITUAL WORLDS

The Popol Vuh shows that by means of the imagination of the Creator God First Father, his son First Lord (or One Hunapu) thrust upward the World Tree at the beginning of time to create a new world. As Douglas Gillette describes this action, "This wondrous Tree had its roots in the Underworld. On the earth plane, it shaped all time and space. Its topmost branches spread into the Overworld where they organized the space-time of the heavens and set the star fields in motion."[3]

Gillette shows—by painted plates that were buried with the dead to help them travel easily in the other worlds—that the Maya believed *the dual spiritual dimensions of the lower and upper worlds completely surround and enfold our world of normal space and time*.[4] As you will see when I discuss sacred postures, for me this idea is real and experiential, and most importantly, all these levels are completely accessible to anyone. During our Pleiadian Agenda Activations, the root system of the Tree is the first and second dimensions, the trunk is the third and fourth, and the branches and leaves are the upper five dimensions. I cannot imagine being in solid reality without orienting myself by this multidimensional enfoldment, because the many worlds offer 90 percent more than what I perceive in the material world. If your mind is open to this possibility, then it is easier to see how the Maya and other people working with sacred trees were able to access all the knowledge in the universe. Members of sacred cultures know a lot, and contemporary Westerners have often forgotten almost everything except what they can see every day; this perceptual loss makes the solid world into an unnecessary and limiting prison. As more people catch on to the way reality is actually constructed, they will be frustrated by a life that has no access to the other realms, and they will begin to switch on multidimensional access just out of sheer boredom.

Beginning with the concept that the upper and lower worlds completely enfold the material world, all sacred traditions find a center in the material world by the four sacred directions. *The World Tree generates the four sacred directions of the solid world*. These directions are not just geographical North, South, East, and West. The sacred view for Native Americans and the Maya is that spiritual qualities arrive in the material world from each direction, and when we "center," we can see and hear "spirit" by reading the information coming in from the directions. Centering means generating the tree in our bodies, and spirit is simply the knowledge that exists in the unseen worlds, which are just as real as the visible one. Simply put, in a specific location, energies that come from the East offer us spiritual direction, from the South nurturance, from the West transformation, and from the North great cosmic knowledge. For example, most of what I know about the Mayan Calendar has come to me from the North. When I sit in ceremony within my altar to the directions, this altar is the center of the ceremony—and for that ceremony, the center of the universe. Sometimes, a sacred temple or a pyramid in a given location is the center where spiritual energy arrives into that place. When I do ceremony, I always set an altar to the four directions and then pray to seven directions. This adds the upper and lower worlds and center, which is my heart united with the hearts of all the other people in the ceremony.

During activations, I always think of my altar as center in that space. In June of 2005, however, after I fully understood Calleman's work, I suddenly realized that *the World Tree is the center during Pleiadian Activations*. This revelation has fundamentally changed my work and my life. Knowing this demands that I teach Calleman's work before I do my own teaching, at least until enough students have studied his work and this book themselves. I am now aware that Pleiadian Activations directly affect the global energy field, which is both an exciting and a challenging realization.

THE WORLD TREE AS THE DRIVER OF EVOLUTION ON EARTH

Calleman's sense of the World Tree is revolutionary, because he has developed a viable theory for how the World Tree drives planetary evolution. Just as a tree's development accelerates—beginning as a tiny seed, generating a trunk and sprouting branches, growing leaves and flowers during its magnificent maturation—the evolutionary force begins in single cells and eventually becomes complex animals; now we as complex humans are expressing this force.

Why humans and not animals? As far as I can tell, my dog Rambo has not changed during the Galactic Underworld, except that he wishes I'd slow down and take him for more walks. The Mechica, the indigenous people of Mexico, say our age is the Age of Flowers because the Tree is flowering now. Speaking of a central World Tree, if there were only individual sacred centers, there would be no evolution, no difference between cultures, and no connection between anything. Maya and Aztec rituals center on the World Tree, which does suggest there is one key center, even though many ceremonies also work with individual centers. This means the whole world is generated, organized, and evolving according to the World Tree.

After assuming the tree to be real, Calleman took a huge critical leap: He wondered where the *geographical center* of the World Tree is located. As far as I know, he is the first person to ask that question, even among indigenous people. As an indigenous person, I have created ceremonies for many years, yet this idea is a truly radical shift that plugs my mind into the planet in an entirely new way. Right now, I really do not comprehend the full implications of this possibility, yet I can feel Calleman is right, so I am just living it out ceremonially through 2011. Whether Calleman is right and whether he has selected the correct center for the World Tree (as well as the *exactly* correct dates for the Nine Underworlds) will be proven or rejected in time, by people either resonating with or rejecting the idea, especially indigenous people. Remember, there are indigenous Scandinavians, Scots, Gauls, Egyptians, Russians, et al. I resonate with the idea of Calleman's center for the World Tree. Since this is a global recognition by people who resonate with Earth, I will do my best to describe his ideas and add my own.

Calleman notes there is a radical difference between the people and cultures of the East and the West. This is obviously true and, of course, it is the cause of tremendous change and instability in the world. As Calleman puts it, "While the East has been dominated by collective structures and has a meditative streak, the West is individualistic, extroverted, and action-oriented."[5] I've also written about how to resolve the differences between East and West. My graduate program was in Creation-Centered Theology directed by the theologian Matthew Fox, who believes that awakening meditative spiritual consciousness in Westerners helps balance and alleviate East–West tension and Western aggression.[6] Our Pleiadian Agenda Activations help balance East–West differences because they not only offer Westerners access to spirit, they also offer Easterners grounding, which they often seriously lack. Still, during my lifetime, tensions between the East and West have intensified, and lately, war and terrorism have engulfed our planet largely because destructive forces (such as arms manufacturers) manipulate peoples' basic perceptual differences.

The violence is not worth it to the people; it wasn't worth it during the Crusades, and the fighting will never be worth it. One of the reasons I am so drawn to Calleman's work is that he realizes that the Maya and Native Americans are deeply spiritual cultures that *geographically* are also Western cultures. In other words, *understanding Native American and Mesoamerican spirituality offers a resolution for East–West conflict.*

THE GEOGRAPHICAL LOCATION OF THE WORLD TREE

To find the location of the World Tree, Calleman looked for the line separating East and West. He chose the midline of the continental mass of Earth from the western tip of Alaska to the eastern tip of Siberia, which is longitude 12° east, hereinafter referred to as the "midline." This is logical and quite visually compelling when you spread out a world map. Calleman notes, however, that physical geography is not the only clue, since this is also a "spiritual geography"—an organizing system influencing human events.[7]

Since the Maya have left us clear records of how the World Tree was

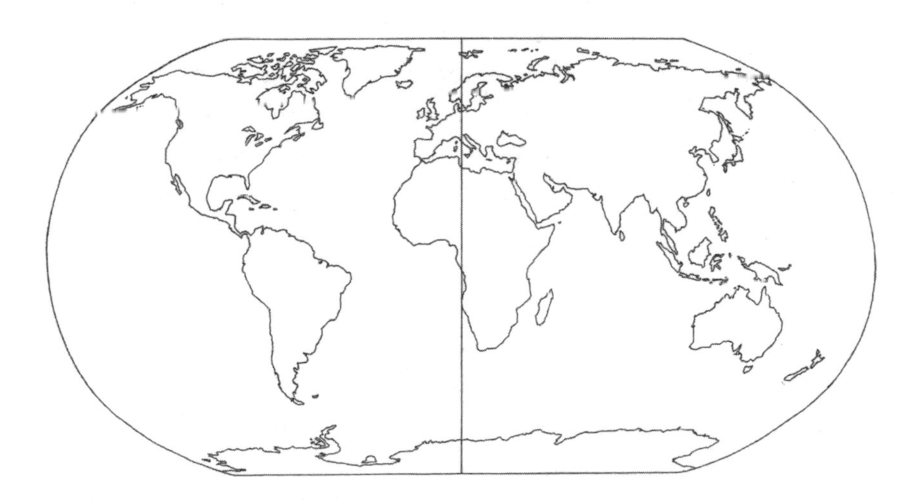

Fig. 5.1. *The latitude 12° east midline of the World Tree.*

created, as well as having said that the Tree generates changes in time, the Mayan Calendar must be a description of this spiritual geography. Therefore, the National Underworld, when the time acceleration was fast enough to generate visible history, is the logical place to look for evidence of the World Tree's influence. If the location of the Tree could be found, then evidence of its influence should be visible as historical movements during the 394-year baktuns of the Long Count. As you will see, the influence of the World Tree is very detectable historically during the National Underworld. This is is an advancement beyond how José Argüelles analyzed global historical changes during the thirteen baktuns of the Long Count in *The Mayan Factor.*[8]

As Calleman considered how human thought might be influenced by the World Tree, he used the Hermetic principle "As above, so below" to find correspondences between the global brain of the planet and its surrounding energy envelopes (above), and the human skull and brain (below).[9] This captured my interest, because in many esoteric traditions there is a literal obsession with the human skull, especially Mayan crystal skulls. In 1990, I did a past-life regression session in which I read the crystal skull of Dzibichaltun described in *The Mind Chronicles.*[10] In the regression, this skull was a record keeper for the Nine Underworlds, which, in terms of Calleman's ideas, is quite compelling. I seemed to have accessed a similar idea about the Nine Underworlds in 1990, but back then I couldn't imagine what it all meant. Also, because I had read the phenomenally brilliant *Earth Ascending* by Argüelles, in which he maps the relationships between Earth's fields and the human brain, I was very open to the possibility of these energetic connections.[11]

Calleman's full theory is called *brain lateralization* and is presented in full detail in his books. Simply put, his idea is that the functioning of the human skull is related to the geographical and energy structures of Earth. As you will see, he may be correct. Also, Argüelles seems to think that this is the case as well, since he wrote the introduction for Calleman's 2004 book. The way this works, according to Calleman, is that, with the midline (longitude 12° east) as the key divider, the Western Hemisphere of the planet parallels the left brain—rational and analytical thought—and the Eastern Hemisphere parallels the right brain—intuitive, spatial, and artistic thought. Accordingly, the West is more action-oriented and rational, and the East is more meditative and intuitive.[12]

By examining history to see what was going on between the East and

West during the National Underworld with the midline as a dividing line, Calleman uncovered a distinctive "wave pattern" of history that first becomes visible during the founding of Rome in 753 BC, the beginning of the Seventh Heaven of the National Underworld. (Remember, there are *thirteen* Heavens consisting of six Nights and seven Days during each Underworld.) Noting the previously mentioned parallels between locations on the planet and the structure of the human skull, Calleman concluded that the planet must have a regulatory center corresponding to that of the human brain—the hypothalamus-hypophysis complex. In reference to the midline, then, he concluded that Italy and Germany represent that planetary regulatory center.[13] These are the areas that dominated history in the West for approximately fifteen hundred years, and their historical patterns feel like a knot in our heads that needs to loosen now. In a moment, we will look at European history starting in 753 BC, but first we need to understand some other geographical ideas.

THE CROSSING OF THE WORLD TREE

The next question that arises is, where is the geographical *crossing* of the World Tree? Calleman notes that this is Gabon in West Central Africa, where the longitude 12° east line crosses the equator. This crossing is very close to the Olduvai Gorge in Tanzania where most of the earliest remains of *Homo erectus* were discovered, which I discussed in chapter 2. In other words, once the evolution of hominids was kicked off during the Tribal Underworld 2 million years ago, according to archaeological and DNA evidence, *hominids began to evolve into erect humans at a spot that was very close to the crossing of the World Tree.* It certainly would be fascinating if the ancient story of Adam and Eve and the fruit plucked from the tree of the knowledge of good and evil in Genesis were a coded reference to the creation of humans near the World Tree.

I bring this up because it would appear from Maya legends that the Classic Maya believed that the World Tree was generated in the imagination of First Father at the beginning of the 5,125-year long Great Cycle or the National Underworld (also the Long Count). Even though this is what the Classic Maya may have thought, there is major evidence the World Tree was influencing planetary evolution well *before* 3115 BC. First, the location of early hominid evolution is close to the Tree. Second, Pangea, the supercontinent that existed before the current continents broke apart because of continental drift, moved out from the crossing point of the midline as it began separating into continents in the middle of the 820-million-year Mammalian Underworld.[14] Indeed, the World Tree is a great mystery that deserves much exploration. Here, we will just have a look at its influence during the National Underworld.

Since the midline goes right through Europe, Calleman's analysis is by necessity Eurocentric, even though his book is about the Maya. Calleman notes that during the period we are going to examine, *the Mayan Calendar was not used in Europe*. About this significant fact he comments, "We know for certain that the movements we have observed are not the results of self-fulfilling prophecies. Instead, they are objective results of the directional winds generated by the World Tree."[15]

Also, regarding the influence of the World Tree in the Americas, American academic dogma about the peopling of the Americas ignores the probable diffusion to the Americas by early people sailing the Atlantic and Pacific. According to the prevailing story, the ancestors of the indigenous Americans ripped across the Bering Straits around twelve thousand years ago and ran all the way down through South America, and then later Columbus discovered America in 1492. So, if movements in the Americas *were* influenced by the winds of the World Tree, this can't be detected, because academics deny diffusion in favor of the migrations over the Bering Straits. Fortunately, we can look at European and eastern European history in light of the midline, because European history is well understood. It always saddens me that history must be so Eurocentric, since the real history of the Americas is so fascinating. But not to worry, all the truth is coming out now!

THE INFLUENCE OF THE WORLD TREE DURING THE NATIONAL UNDERWORLD

We see the first action on the midline at the beginning of the Seventh Heaven (749 BC), when Rome was founded around 753 BC. Once a historical culture became activated on the midline, an interesting pattern emerged during the rest of the National Underworld. Before this time, civilization developed in the Fertile Crescent, and then gradually people migrated west. By 749 BC, there was sufficient settlement near the mid-line to generate a pattern, as history in that area became visible.

Regarding the World Tree's influence around the midline during much earlier times, there was a great amount of activity near the midline prior to twelve thousand years ago, for example, with the Magdalenian culture. During the Seventh Heaven, Rome was founded on the midline, where it garnered significant power and influence, which still exists today. Noting from chapter 2 that the Nine Underworlds are divided into six Nights and seven Days of the Thirteen Heavens, Calleman found the geographical pattern to be that *during odd-numbered Heavens, history is generated on the midline, and during even-numbered Heavens, history is drawn to the midline far from the East.*[16] As you will see, this pattern is quite dramatic and specific. Also, *as history progresses, events tend to move up the midline.* We will look at this pattern very briefly, and I highly recommend Calleman's deeper discussion called "Winds of History."[17] Here I will just briefly summarize the pattern.

Rome was founded at the beginning of the Seventh Heaven, and then during the opening katun of the Eighth Heaven (355–335 BC), the Persians under Ataxerxes III moved west, conquering Egypt, Asia Minor, and the Athenians. During the beginning katun of the Ninth Heaven (AD 40–60), Rome on the midline activated again, and the Romans conquered England, Wales, and parts of Germany to the north, Morocco and Algeria to the west, and Bulgaria to the east. During the beginning katun of the Tenth Heaven (AD 434–454), the central Asian Huns under Attila broke down the Roman Empire, throwing Europe into the Dark

Heaven Number	Time Period	Movements from the Central Line of 12th Longitude	Movements from the East
7	749–729 BC	Settling of Rome	
8	355–335 BC		Persians go West
9	AD 40–60	Expansion of the Roman Empire	
10	AD 434–454		Huns go West
11	AD 829–849	Raids of the Vikings	
12	AD 1223–1243		Mongol storm
13	AD 1617–1637	Sweden	

Fig. 5.2. *Violent migratory movements away from and toward the planetary midline. (Illustration from Calleman,* The Mayan Calendar and the Transformation of Consciousness.*)*

Ages. At the beginning katun of the Eleventh Heaven (AD 829–849), the midline activated again, but further north this time as the Vikings mysteriously launched serpent-adorned ships and blasted off from the midline, raiding the British Isles to the west and Russia to the east. Also during this time, the modern nation-states of France and Germany emerged from Charlemagne's empire.[18]

During the opening katun of the Twelfth Heaven (AD 1223–1243), the Mongols under Genghis Khan roared out of central Asia, conquering everything before them, creating a "unified field of the Eurasian continent" and "the largest empire in human history."[19] Lastly, during the beginning katun of the Thirteenth Heaven (AD 1617–1637), the Thirty Years War between Protestants and Catholics ignited, and Sweden temporarily became a major power (energy moving up the midline) during this struggle, which ended up weakening the papacy for the first time in 1,500 years, allowing space for the Protestant Reformation.[20]

Alas, as the papacy fought to retain its influence, it condemned great new scientists such as Galileo, whose struggle with the Vatican spans the first katun of the Thirteenth Heaven. Catholic credibility was gradually eroded over the next few hundred years because of the way Galileo was treated by the Vatican, and then the world realized Galileo was right. Also, during this katun, in 1620 the Pilgrims went to America, which eventually became an extension of England in the European power game. We are now living out the last katun of the Thirteenth Heaven, which is also Day Seven of the National Underworld, the fruition of 5,125 years of history.

The midline still generates great energy at the end of this cycle, which suggests the European Union (EU) will get stronger as America weakens, although by admitting so many eastern European nations, the EU may have compromised its strength. I say this to point out that knowing the pattern just described could be very useful politically and financially, and I will investigate questions like this in the following chapters on the Galactic Underworld. Next, we need to get into the deeper questions that emerge out of the pattern just articulated.

THE COLLECTIVE DRAMA AND THE INVISIBLE CROSS

Considering the historical waveform—the winds of history—only invisible boundaries such as the World Tree can explain such a dominant pattern. Calleman notes, "This Cross of creation field boundaries introduces a creative tension along the lines where it is introduced, and this creative tension results in, among other things, migratory movements of people away from these lines."[21]

As we've said, this is a wave pattern of history that energizes flows of dualities and unities in the human field, inspiring mass movements of people; in other words, "the cosmic drama which appears to be enacted by the forces of the West and the East, male and female, yin and yang, or Light and Darkness."[22] The idea of a drama involving male and female forces acknowledges that most ancient sacred cultures recognized the interplay between unity and duality. In the more recent times of the National Underworld, however, people are unconsciously swept back and forth by the midline, like trash blown by the wind in an aging city.

In my work with nine dimensions of consciousness during the Pleiadian Agenda Activations, the third dimension (3D) is the solid world, and the collective mind (where the World Tree would influence people) is the fourth dimension (4D). Just as the midline is invisible, so is 4D invisible, yet 4D controls the collective mind of the people. For example, consider the conflict that Judeo-Christianity plays out with Islam. Most media-led Americans continuously run tapes in their minds about Islam, terrorism, and the Crusades; for them, the world is divided between good and evil. Considering Calleman's historical analysis of the midline, the West has an instinctual fear of the East based on the arrival of the Persians under Ataxerxes III, the Huns under Attila, and the Mongols under Genghis Khan; for Europeans around the midline, monsters just suddenly appeared. Those in the East would have been aware of the West's accretion of wealth and power during the odd-numbered Heavens. During the even-numbered Heavens, they would have suddenly felt compelled to move west to raid and conquer.

Our brains and nervous systems seem to be hard-wired into the patterns created by the World Tree. *This means the human collective mind is globally wired into the periodicity of the cycles of the World Tree, which are delineated by the Mayan Calendar.* This astonishing idea may explain the potent grip of 4D that we have detected within the minds of our students during activations. And like all patterns that grip the human mind, if we can identify the pattern, see how it affects the planet, and choose to create in a new way, our species could get off this treadmill of mindless duality and violence. Most importantly, surely the end of the

Mayan Calendar must mean this midline influence is ending; this would be great news in light of the patterns I've just described.

HOLOGRAPHIC RESONANCE BETWEEN THE HUMAN BRAIN AND EARTH

During Pleiadian Agenda Activations since 1995, our students have been exploring the roots of the World Tree by traveling into Earth's core and retrieving information to activate our consciousness. As I said in *Alchemy of Nine Dimensions*, "The incredible truth is that we vibrate with the pulse of Earth, which aligns us with all other beings—including light—in the ladder or chain of existence. This is the vertical axis of consciousness."[23] The system of nine dimensions we work with really does describe the different aspects of human consciousness, but we've often been stumped by the control of 4D in our lives. It turns out that Calleman's concept of the functioning of the World Tree explains how this works, especially considering that our brain waves are entrained by Earth's vibratory patterns. That is, our brains register the vibrations of 4D whether we are conscious of this or not.

The iron-core crystal in the center of Earth vibrates at 40 Hertz (Hz), 40 pulses per second. When the mind is most active and creative, it pulsates with the core at 40 Hz—beta waves—which means *we are resonating with the Earth's inner core*. When the mind is relaxed and meditative, it vibrates with Earth's inner spheres, those from the core to the crust,

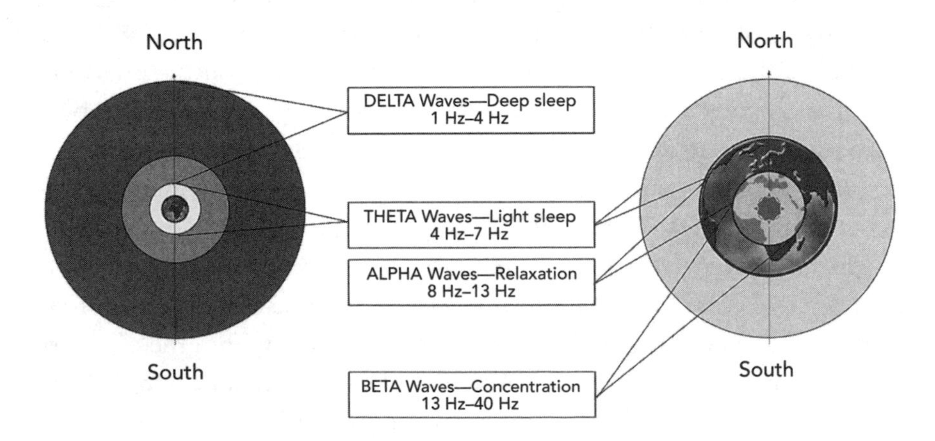

Fig. 5.3. *The human brain's resonance with Earth. (Illustration from Calleman,* The Mayan Calendar and the Transformation of Consciousness.*)*

which pulse progressively more slowly from 40 down to 7 Hz—beta through alpha—meaning *we are resonating with the inner Earth.*

Our brains vibrate in alpha at 13 to 8 Hz, so our resonance with the near inner Earth is in both beta and alpha waves. The Earth's crust vibrates at around 7.5 Hz, the transition between alpha and theta waves. When we are drowsy or in light sleep and our consciousness is beginning to move out into the atmosphere, our minds pulse with theta waves at 7 to 4 Hz. Earth's inner Van Allen belt, the inner belt of radiation that contains charged particles that are held there by the Earth's magnetic field, vibrates at around 4 Hz. The range between Earth's surface and the inner Van Allen belt is 7 to 4 Hz, which means when our minds pulse in theta, *we are resonating with the atmosphere and the inner Van Allen belt.* When we are in deep sleep, we are in the delta range in our minds at 4 to 1 Hz, which means we are resonating with the outer Van Allen belt and out to the magnetopause, the interface between Earth's electromagnetic fields and the solar system.[24]

In other words, when we are in the deepest states of concentration, we are hooked right into the core of Earth's central intelligence, and when we let go, relax, and fall asleep, our minds move out of Earth's fields. When we are thinking the most intensely and having creative flashes of genius, we are vibrating with the iron-core crystal, and when we move out into space, we are leaving our bodies and Earth and moving into the celestial realms. In my nine-dimensional system, the first dimension (1D) is the iron-core crystal in the center of Earth, 2D is the outer and inner core, 3D is the crust, and 4D is Earth's atmosphere extending all the way through the inner and outer Van Allen belts and out to the magnetopause. The higher five dimensions are beyond the magnetopause. Very adept students know where they are in the dimensions just by reading their own brain frequencies. Calleman notes, "The concordance between the traditional ranges for the frequencies of different types of brain waves and the radiuses of earth's spheres is remarkable."[25] Clearly, our brains are transducers of all the elements of Earth and the cosmos, and once people realize this, they will stop destroying their habitat. For example, the fields our brains can read in the atmosphere are being altered by microwave frequencies created by cell towers and microwave ovens, and what these alterations might be doing to planetary biology will soon have to be taken very seriously.

Calleman speculates on what the resonance between the human

beta-wave frequency and Earth's inner core at 40 Hz means regarding the perspective of the Mayan Calendar. This question is most fascinating in light of my nine-dimensional theory, because the central reason for each activation taking place is to get into the nature of consciousness in Earth's core and the Galactic Center. Science posits that a dense, hot, pressured crystalline structure is in the center of Earth.[26] Because of detected irregularities, this mass is not a perfect sphere, and Calleman speculates that, since the ancients often depicted the World Mountain as a pyramid, the inner core may have an octahedral structure, which could generate a midline.[27] This is a very important point, because, according to the Maya, the World Tree is anchored in the World Mountain in the center of Earth, and it projects perpendicular planes or branches out to the surface.[28] Calleman conceives of the midline as one of the branches of the World Tree.[29]

Most importantly here, Calleman suggests that models of pyramids might be the source of the term *Underworlds* for the nine levels of creation. As he puts it, "These Underworlds, as the terrace-like pyramidal models of the World Mountain seem to imply, might originate in the crystalline structure of the core."[30] And he wonders, do the "Nine Underworlds correspond to nine sequentially activated layers of iron crystals in the Earth's inner core?"[31] I have to add his thoughts, because this is exactly what many of our students discover in the iron-core crystal in the center of Earth. After all, the dimensions arise out of the iron-core crystal on the vertical axis of consciousness, thus the iron core contains the records of all the dimensions.

During activations, when we go into Earth's core with prayers and respect, the most consistent and enduring experience people have is of a timeless Eden, a creation place that contains all species of the plants and animals that have ever appeared on Earth's surface. Some people seem to be traveling in the Cellular Underworld or in the early stages of the Mammalian Underworld, and many people have astounding revelations about the species and their changing habitats.

The refrain is usually that all species exist eternally in Earth's core crystal and could return to inhabit the surface if a habitat existed in Earth's material realm to support them. In other words, the *habitat on the surface determines what is actually created,* while all the aspects of creation exist in potential form in the center of Earth. As we have seen, deep thought and intense concentration while we are in beta frequency puts us in resonance with Earth's core. *Beta is a very high frequency that can be thought of as enlightenment.* Calleman says, "Raising our frequencies and awakening our minds through resonance with the earth's inner core are tantamount to following a path toward enlightenment."[32]

ACTIVE MEDITATION

When I was a kid, I was very fond of cowboy shows; once I even wrote a love letter to the cowboy star Roy Rogers. His wife, Dale Evans, sent me a signed picture of Roy and said she didn't mind my feelings, which greatly relieved my guilty mind. As glamorous as Roy and the other cowboys were, I see now that the one who actually influenced me was Tonto, the Lone Ranger's sidekick. Tonto just sat on his horse and repeatedly grunted, "How?" and again, "How?" Early on, I adopted the idea that the only thing that matters is asking "How?"

I like activations for seeking enlightenment, and sacred postures for traveling into the other worlds, because these things work! I end this chapter with a few speculations on valuable practicalities, and some thoughts about how to use postures to work with the Sacred Tree. As I get into it, remember that Felicitas Goodman—the modern discoverer of ritual trance and sacred postures—noticed that during trance under rhythmic stimulation the brain goes into the theta range, as discussed at the end of chapter 3. That is, when we are in trance and traveling in the World Tree, our minds are moving out into the upper atmosphere and the inner Van Allen belt. In other words, regarding ecstatic trance, the alternate reality is accessible to us by leaving Earth. In my experience this is true, and this is why our trance sessions are never longer than fifteen minutes—yet sometimes feel like hours.

From a practical point of view, you can make an appointment with yourself to travel in the World Tree, which exists in the alternate reality, the invisible realm that co-exists with the material realm. Here I will label this work as a form of *active meditation.* You can create a four-directional meditation space in your house or temple and go into deep states of meditation. Then, using your own body as the World Tree, you can travel up or down the multidimensional vertical axis.

The first thing you might ask is, How does this differ from traditional meditation, such as Eastern-inspired states of stillness? The two forms differ absolutely in their goals, because World Tree meditation is very active and filled with content, while Eastern sitting meditation empties one of thought. One technique is not better than the other, yet in my experience almost all people born in Western cultures can achieve active meditation, yet they struggle with Eastern forms. I have noticed in our activations that the opposite is also true: Practitioners of the Eastern forms suffer while attempting to engage in active meditation. Active meditation has been relatively unavailable since the Middle Ages, when nuns and monks practiced it daily by chanting, painting illuminated manuscripts, walking in sacred gardens and labyrinths, and reading sacred scripture. During these processes, people were intentionally modifying their brain frequencies to access altered states. Perhaps these examples represent the extreme value of active meditation for us today: Active meditation is a contemplative technique that provides Westerners with a method to commune with the Divine.

Eastern spiritual teachers also use active meditation techniques such as chanting and yoga, which are similar to the methods favored in the West during the Middle Ages. But when Eastern meditation techniques first came to the West in the 1970s, silent and passive meditation was primarily offered. From the Eastern perspective, Westerners are spiritually undeveloped and often a little crazy. I do not disagree with this assessment, because the intellectual databank (of Western culture) that developed during the last four hundred years—Cartesian dualism—offers only half of life. Regarding the various forms of Eastern meditation, teachers were wise to suggest people should empty their minds, since their minds were filled with junk!

It is fair to say that the same situation has also existed in the East, since access to the deep truth (such as the meaning and antiquity of the Vedas) was lost until very recently, as discussed in chapter 4 regarding B. G. Sidharth's research.[33] Meanwhile, now that the West is recovering its own spiritual story by means of new-paradigm research, active contemplation is a very potent way for us to achieve spiritual awareness. For example, by actively contemplating the cycles of time, we can move right into the ninth dimension—the Galactic Center—the location of the highest states of consciousness that exist. This is contemplation of eternal truth—samadhi. And, speaking of brain waves, the center of the Galaxy pulses with very high-frequency gamma waves, and you have to wonder how that affects our minds.

The practice of using sacred postures is a great big How! But to select this wonderful opportunity, you have to attain a completely new level of respect for your right to explore the universe. Both Eastern and Western religious systems have adopted many proscriptions against the individual's right to knowledge, and now they all have to go. Whether it is the idea that you need a guru to find enlightenment or the idea that personal knowledge is the work of the devil, all must go. Now during the Galactic Underworld, all the knowledge gained during the previous Underworlds has sped up so fast that all this information is pouring into the human mind faster than we can process it. Sacred postures were used during the Regional and National Underworlds; thus, they are special keys into various levels of consciousness.

SACRED POSTURES AND THE ALTERNATE REALITY

At the Cuyamungue Institute, Dr. Felicitas Goodman's anthropological institute in New Mexico, we continue to find ancient postures, yet the categories we work with probably will not change.[34] We have found many postures, some of which assist us with healing, and others that allow us to divine knowledge about things we wish to understand. We work with postures that help us to deconstruct and regenerate ourselves by metamorphosis, as well as with postures for spirit journeys into the lower, middle, and upper worlds. Some are initiation postures, while others assist us with issues of death and rebirth. Finally, there are postures to access living myths so we can see the spiritual aspects of our lives, and postures of celebration to honor the joy in our lives. Anytime we choose, we can select the spiritual state we want to access, and then consult the wisdom available in the alternate reality. Just like a monk or nun who exists in a spiritual state most of the time because of constant devotional practices, we can be plugged into all levels just by regularly visiting the alternate reality.

Much of the day for a monk or nun is mundane. Their daily attunement with God, however, spiritualizes the mundane world. Everyday life is spiritualized by visiting the alternate reality often enough to remain consciously enfolded within it. Additionally, life in the alternate reality is astonishingly rich for us because our ancestors explored it for many thousands of years, and they still live there, waiting for us to return. This databank may be the richest one for modern humans, because it is entirely free of religious manipulation and coercion.

Some may wonder why I also work with Pleiadian Agenda Activations.

We do activations because they are large-group experiences that create a weaving of worlds for many people. Regarding group experiences and ecstatic trance, at Cuyamungue we also conduct masked trance dances, which are very similar to indigenous dances because the spirits come through us as dancers, and together we create a weaving of worlds. Usually over a whole week, the dancers make their own costumes that completely cover their bodies. We are in trance two or three times a day to divine the animal we will be and to understand and create the dance the spirits are calling for in that moment. During this time, we are not in ordinary reality; we are in the alternate reality. During a Pleiadian Agenda Activation, we travel as a group into nine dimensions, while we are each in our bodies attuning to our own brain-wave frequencies. One of the most fascinating things about the activations is that we enter each dimension in a sacred way. When students experience the third dimension—linear space and time—in the sacred way, they discover the sacred Earth, which forever changes them. They learn how to detect other worlds merging with the solid world, and their edges between realities are softer after that. When we experience a masked trance dance, living in the alternate reality with the spirits for a whole week, ordinary reality never looks the same again.

I'd also like to introduce the idea that reading is a sacred practice that can activate the beta frequencies in our brains. Western education and culture has been crammed full of false data and untruthful ideas for the last 400 years, even the last 1,500 years. During this period, only those who have had access to accurate central databanks, such as advanced mathematicians, physicists, and artists, have been able consistently to experience thought in the beta frequency. Now that new-paradigm writers are offering correct information again—information that once existed in thousands of texts that were destroyed in the Alexandrian Library and in Mesoamerica—many people are having brain orgasms, the Aha! experience, which is high beta thinking. I end this chapter asking you to honor the fact that a brain orgasm with a great book is the same as samadhi in the ashram; enlightenment is simply truth and clarity with pure light.

6

THE GALACTIC UNDERWORLD AND TIME ACCELERATION

THE GALACTIC UNDERWORLD AND AMERICA AS GLOBAL EMPIRE

The moment has arrived to examine the nature of time acceleration during the Galactic Underworld, which opened on January 5, 1999. First, let me be clear about how this system works. According to Calleman, the speed of evolution during the Galactic Underworld increased by a factor of twenty over the Planetary Underworld (AD 1755–2011), which increased by twenty over the National Underworld (3115 BC–AD 2011).

Hereinafter, I will simply use the terms *National, Planetary,* and *Galactic* for the Sixth, Seventh, and Eighth Underworlds. Remember that the thirteen stages of the Nine Underworlds are the Heavens, and they also have different Maya numerical names such as *baktuns, katuns,* and *tuns*. During the National, each one of the thirteen baktuns is about 394 years long; during the Planetary, each katun is about 19.7 years long; and during the Galactic, each tun or divine year is 360 days long; yet the evolutionary processes that develop during each one of these Heavens are of *increasing global intensity*. In other words, evolutionary processes that developed (and are still developing, because all Nine Underworlds are stacked on each other) over 394 years, then 19.7 years, are now developing via Days and Nights of less than one year (360 days)! Also, *all the multiples of tuns times twenty going back 16.4 billion years have sped up into just thirteen tuns during the Galactic!*

As I consider the time-acceleration factor during the Galactic, I will look back to the corresponding and slower baktuns of the National and the katuns of the Planetary to seek similar processes that were unfolding. So that you can understand why everything seems to be going faster now, my goal in this chapter will be to give you a tangible sense of the intensity of the Galactic speedup by relating current events to corresponding trends during previous Underworlds.

Since it is impossible to cover in a few pages all major world events since 1999, I have decided to work with a very pointed theme that seems to be having a global impact as the Galactic unfolds. My star player of the global Galactic drama is America as Empire, which is also the title of an excellent book by the American author Jim Garrison, who adopts a view of American politics that is similar to my own.[1] Examining America's role as the global empire in light of Galactic time acceleration adds new depth to this plucky American quest. Simply put, the United States under the aegis of the current Bush administration—the "Bushites"—seized the powers of the time-acceleration factor in 2000, the New Millennium. They adroitly fashioned the unique creative potential of this moment in time into a program for controlling the world.

These clear intentions created instant success, and I have often wondered if they know all about the Mayan Calendar. If you think such a possibility is too far out, notice that the Bushites have been performing like an excellently coached football team directing a runaway game in which the opposing team can't even pass! By passing the ball perfectly and running quickly into the end zones, they have been systematically reorganizing all the governmental systems. As for global leaders who don't cooperate with their plans, the Bushites just pop new rabbits out of the hat every day before anyone can figure out what went on the day before; *the Bushites are having a ball surfing the Galactic time wave!*

This chapter will also occasionally integrate astrological patterns with the Thirteen Heavens of the Galactic, since the astrology of the whole cycle is exceedingly auspicious. The grand finale of a great evolutionary symphony can be found in this planetary ballet, and these astrological cycles are described in detail in appendix B. A formative astrological conjunction—when one planet meets another one in the sky—occurred in 2000, which enabled the Bushites to seize total control. I am sure the Bushite astrologers used it as a guide, and if you doubt this possibility, remember that Nancy and Ronald Reagan's use of astrologers became public knowledge in the mid-1980s. This conjunction was the meeting of Jupiter and Saturn in May of 2000, an aspect that occurs every twenty years. This conjunction—the trigon—is the most accurate astrological tool for tracking political cycles.

For example, since 1840, every American president in office when this conjunction occurred died before leaving office—except Bill Clinton, who, we might say, died of embarrassment over his sexual peccadilloes. FDR and William Harding died naturally, while Abraham Lincoln, William McKinley, and John Kennedy were assassinated, and an assassination attempt was made on Reagan (which may be why his wife sought out the astrologers!). Generally speaking, Saturn rules structure and Jupiter rules money, so if under a trigon a group creates structures that control economies, they win!

The term *Bushites* simply refers to the neoconservative, fundamentalist Judeo-Christian members of the Bush administration (hereinafter the *neocons*). Also, I use the term Bushites to point out that they will not be going away once George Junior finishes his second term. This administration has successfully put in place systemic changes that will be in effect in the United States until AD 2020 (the next Jupiter conjunct Saturn), such as Homeland Security, the taking over of airline security systems, and the reorganization of the CIA and FBI. Unless the American people throw out the Bushites and purge these systems, they are in control. Also, the political formation of America actually began during the first katun (1755–1775) of the Planetary Underworld speedup; thus the United States seized the agenda of the Planetary.

Systemic reorganization of all the levels of government and the military is always the goal of fascist regimes. What do I mean by "fascist"? Fascist systems seek to instigate the total control of individuals by corporations that are owned by the friends of the current leaders. Of course, corporate power has been the goal of U.S. systems since at least the 1950s. However, now fascism is blatantly taking over the United States under the Bushites. In other words, the leaders of the United States use the country's economy and powers to advance the private agendas of certain corporations, in this case the pharmaceuticals, weapons manufacturers, and corporate nation-builders, such as Halliburton in Iraq. We will move back to the beginning of 1999 to explore this theme and trace the nature of time acceleration that is now twenty times faster than it is from 1755 to 2011. Of course, we will also look for completely new themes emerging from the Galactic Underworld speedup.

DAY ONE OF THE GALACTIC UNDERWORLD: JANUARY 5, 1999–DECEMBER 30, 1999

Day One of the Galactic—sowing the seeds—must be examined in detail to detect the new agendas that will unfold all the way through the Thirteen Heavens.

During 1999, there was a major shift in people's lives due to the impact of a universal Internet and e-mail. Suddenly, people who were not computer savvy struggled in the working world as almost all businesses went digital, and young people who were comfortable with computers developed an edge that made mature managers and owners very nervous. E-mail emerged as an eventual necessity, and it was obvious that those who didn't adopt e-mail were going to be tomorrow's dinosaurs. Professional researchers swarmed online, and cell phones were instantly adopted by businesses, which put employees on the job fulltime except

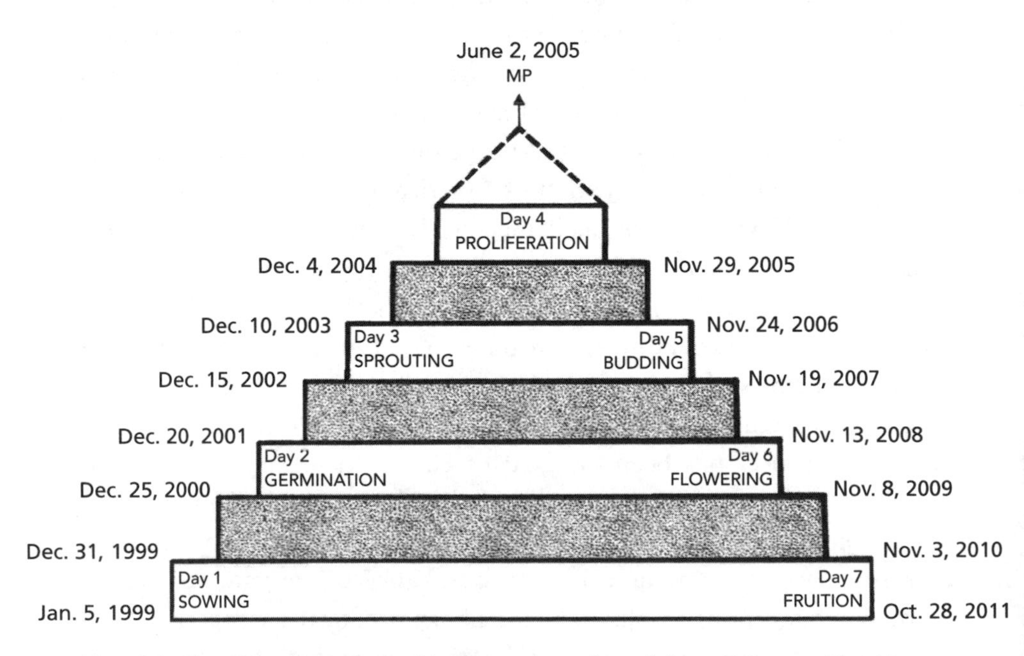

Fig. 6.1. *The Galactic Underworld. (Illustration adapted from Calleman,* The Mayan Calendar and the Transformation of Consciousness.*)*

when they slept. More personal ways of communicating, such as by telephone and "snail mail," were devalued, and people e-mailed fellow workers instead of walking down the hall to ask a question or chat; time sped up radically via e-mail, Internet, and the first cell phones. Do you remember this distorted time speedup?

This phase was very stressful for people, especially for older people. At this time, and based on old dating systems, the computer makers, the government, and the media publicized a computer scare called "Y2K" that was due to occur at the end of 1999. This weird hype, in which the turn of the millennium was predicted to cause havoc with computer systems not designed to handle date stamps that didn't begin with "19," made people feel dependent on computers, and it convinced many people to "upgrade"; computer suppliers were laughing all the way to the bank. As it turned out, computer programmers worked frantically to devise workarounds and patches, and the disaster was averted, but people had internalized the fact that their world was inevitably becoming computerized. *Computers were necessary for survival!* The U.S. economy was in fantastic condition because of the information technology (IT) wave, and the fact that then-President Clinton had greatly reduced the deficit. Worldwide, people were involved in a lot of stock market speculation, mostly in the new technologies, which was reaping huge taxes for the United States. During 1999, the U.S. economy was like a sexually active twenty-year-old ready for more fun, while the president seemed to think *he* was still twenty years old.

Meanwhile, there was a financial shadow (for the United States) lurking in the world: In January 1999, when the Galactic acceleration gas pedal hit the floor, the euro was implemented by the nascent European Union. This new currency signaled possible future challenges to the value of the almighty dollar. In retrospect, we know that the euro was a challenge to the dollar, and we can see that this was a signal for other countries in the world to draw into groups that would serve their own territorial interests. For example, as I write these words in 2006, European and South American countries are consolidating common interests in their hemispheres to handle the threat of reckless American dominance, which is making the Middle East into a powder keg. Countries such as Iran in the Middle East and China in the Far East are building power and alliances at least partly in reaction to U.S. dominance. Also during 1999, a great dissatisfaction was building in the United States (which did not come to light until 2000) among conservatives and religious fundamentalists. They were angry about the personal freedoms (such as gay liberation and safe abortions) that many Americans had endorsed and were getting used to.

Looking back to Day One of the National Underworld (3115–2721 BC) for similar themes, complex temple-city cultures with writing systems and hierarchical cultures suddenly cropped up in many places on the planet. The process that is unfolding during the whole National Underworld is the development of civilization and communication, which is in its most advanced stages now. Looking back to Day One of the Planetary Underworld (AD 1755–1775), early industry suddenly transformed the agrarian, rural world that produced food for the cities where organized economies and trade existed. Many people were pulled out of bucolic country living to work like slaves in the new factories in these cities, and early manufacturing spawned global networks, which are culminating now. In 1755, the American Colonies began seeking independence from European control, and that led to the Second Continental Congress ratifying the Declaration of Independence in 1776.

The United States defined itself during Day One of the Planetary, so it is not surprising that it is so dominant during Day One of the Galactic. Just like Day One of the National, Day One of the Planetary forced nascent desires to the surface that accelerated reality whether people liked it or not, and it is exactly the same now. During Day One of the Galactic, the IT revolution was shocking for all but a few people in the know, like Bill Gates of Microsoft and Steve Jobs of Apple. Suddenly in 1999, we all knew the old world of seeking material comfort and security was ending, and boundaries were eroding everywhere; *every day felt like a whirlwind.*

Sometimes I long for the Victorian Era, just as people during the early National longed for Eden. Starting in 1999, we realized we'd be forced to connect with other people in rapid and new ways—again, whether we liked it or not. Insidiously, advertising became more dominant than ever before, and pharmaceutical corporations in particular began spending massive sums on advertising. This errant turn has fundamentally changed medicine from caring for people to using people's health for corporate goals. I mention this because this threat to well-being has become a global crisis, yet we are only just past the midpoint (June 2, 2005) of the Galactic Underworld. Health is so fundamental to human happiness that

I will trace pharmaceutical oppression though the Galactic, especially since pharmaceuticals are a key linchpin of Bushite economics.

In early January 1999, demonstrations against the World Trade Organization (WTO) exploded in Seattle, Washington, and blossomed into a youth movement against the WTO, the International Monetary Fund (IMF), and the World Bank. In the short run, this was an underground movement, partially because so many demonstrators were thrown in jail without access to a lawyer, but primarily because of the Bushites' obvious intention to eliminate civil liberties. Counterbalancing the anti-WTO movement, young Christian conservatives were working very hard at the grassroots level to attain more power and recognition within the American political system.

It is important to recognize the spectrum of viewpoints in the younger generations in 1999, since the world will eventually belong to them. Spiritual values are building all the way through the Galactic, as both youth movements struggle to break down materialistic corporate control. Perennial and fundamentalist spirituality will dominate in 2011, and these seemingly opposing forces will unite.

Lastly, we can now see that very unpleasant warlike tendencies are showing up during the Days of the Galactic, such as Bush's 2003 invasion of Iraq during Day Three. During 1999, the NATO/Western war for Kosovo was instigated to curb Serbian power. As you'll recall from chapter 2, new creation occurs during the Days of the Underworlds, and the integration of that growth occurs during the Nights.

Since using war to push agendas has been a dominant practice during the Days of both the National and Planetary Underworlds, this is likely to continue during the Days of the Galactic, which is the case so far. The Balkans suddenly became the focus during the last year of Clinton's reign in 1999, which otherwise had been quite peaceful since 1992. Can you remember how surprised you were about that? I will trace war as a knee-jerk reaction to the need to get things done through the Days of the Galactic.

Wars are what the militaristic economy of the United States requires in order to play empire. However, for the sake of planetary survival, those who can see that wars are actually only about money need to work to ensure that the period of the Galactic Underworld is the last gasp of this hideous human tool. At the end of chapter 7, I imagine how our world could go beyond war during the end of the Galactic Underworld.

NIGHT ONE OF THE GALACTIC UNDERWORLD: DECEMBER 31, 1999–DECEMBER 24, 2000

Night One of the Galactic was a period of integration after the astonishing acceleration during 1999. The world woke up after New Year's parties celebrating the New Millennium, and the lights and computers were still on! The world was saved from the Y2K doomsday scenario! At this time, only the neocons and the global elite (a worldwide consortium of bankers and military-industrial leaders) knew that George W. Bush would be pulled out of the hat to lead the new global empire. The empire required royalty, and King Daddy Bush had seeded a royal line—George, Neil, and Jeb.

The environment was perfect for a blue-blood takeover, since the uncouth Bill Clinton had unzipped his pants at all the wrong times and in all the wrong places. A lot of people were nervous because the NASDAQ dropped 600 points on April 4, 2000. After twenty years of rising prosperity and a healthy economy (the previous Jupiter/Saturn cycle), most intelligent people knew it was time for a crash. However, there was speculation frenzy in the new and confusing IT economy because people were operating on greed. One moment from that period is indelibly imprinted in my memory.

While listening to National Public Radio during the spring of 2000, I was startled to hear the barely intelligible stammering of an extraordinarily ignorant politician thrashing his way through an interview. His annoying nasal voice was cocky, pugnacious, and arrogant. I struggled to hear what this odd person was trying to say. When the end of the interview finally came, I learned that the speaker was George W. Bush, son of George Herbert Walker Bush! An idiot, who sounded like he was swishing marbles in his mouth when he spoke, was being jacked up to be president of the United States!

The presidential campaign during 2000 was bizarre, since the current lame duck, Bill Clinton, was up for impeachment and was being vilified for his sexual improprieties. Media attention was on Monica Lewinsky's dress and poor Hillary Clinton's marital problems—everything but the critical importance of the 2000 election. This election was major, since Jupiter and Saturn came together in the sky in May of 2000. The hyperactive IT economy was building and crashing, and the American people were nervous, judgmental, righteous, and polarized.

Ultimately, many people believe the outcome of the 2000 election was fraudulent because of computerized voting machines. Diebold Corporation manufactured the voting machines, and its CEO was on record as having personally committed to deliver Ohio's electoral votes in 2000 to Bush. Clearly, election fraud—the bane of democracies—is a real problem in the United States.[2] The antidote to computerized election fraud is a paper trail, delivered by virtually all other machines (such as ATMs) that Diebold makes. Under these circumstances, the Supreme Court ruled against manual recounts to resolve the disputed election, and George W. Bush seized the mantle of control for the neocons for the next twenty years. This bold model—the United States as global empire for God—was set firmly in place during Night Two.

As we proceed into Day Two, I will not look back to Night One (or any other Nights) during the National and Planetary Underworlds, since the Nights are the integration of the Days, where the action is. So, let us see what emerges during Day Two of the Galactic. Do you remember that crazy year when the media was yapping about big Bill's sexual escapades, while a sneaky, snake-nosed hegemonist was stealing onto the world's stage armed with his daddy's personal cadre of Blue Meanies, the villains the Beatles portrayed in "Yellow Submarine"?

DAY TWO OF THE GALACTIC UNDERWORLD: DECEMBER 25, 2000–DECEMBER 19, 2001

Day Two of the Galactic—germination of the seeds—opened during a period when the national election in the United States was being challenged in the courts. Protests against the WTO and materialism were intensifying, and the environmental movement was growing.

Behind the scenes, George W. Bush was busy working on forming his cabinet and choosing other appointees, since he knew he was a shoe-in. Once Gore conceded (supposedly to offer the United States stability in government), Bush put his chain of command on the tracks and roared ahead like a high-speed train. The public was swept along by glowing reports in the media about the president's great team, while nearly daily bombings in Iraq's no-fly zone were going on, and sanctions against aid to Iraq were causing the deaths of many thousands of Iraqi civilians. In the media, Americans heard constant fear-mongering about Saddam n, while inside Bush's cabinet, from the very first day, plans were ...ormulated to invade Iraq.

It is common knowledge that Bush planned to go into Iraq from the beginning. We know this because Paul O'Neill, the Bush-selected U.S. treasury secretary, resigned his cabinet position in January of 2003 and wrote a tell-all book with the journalist Ron Suskind. *The Price of Loyalty* covers the first two years of Bush's reign, when O'Neill was worried about Bush's lack of engagement on domestic issues while he planned for "regime change" for Iraq.[3]

Utilizing nineteen thousand internal documents that O'Neill provided to Suskind, plus his own detailed testimony of what went on in the cabinet and the National Security Council from January of 2001 through January of 2003, O'Neill exposed Bush's real agenda from the beginning. As soon as O'Neill's book was published in January 2004 (it sold five hundred thousand hardback editions and is in paperback now), many Americans got the truth; certainly everyone in the government read it carefully. In other words, Washington knows the emperor has no clothes!

The Price of Loyalty totally exposes what went on inside the Bush cabal during 2001, the germination phase of the Galactic, which is why I chose the theme of America as global empire. We will follow this potent American political quest through the Days and Nights of the Galactic, since there is no excuse for intelligent people to not see what is going on here. Men like Paul O'Neill risk their lives to inform the public. Will the American people under Bush act as the German people did under Hitler? Since the Galactic acceleration is a spiritual awakening, this is not a likely outcome.

Of course, the big news for America in Day Two is the attacks on the World Trade Center in New York City and the Pentagon in Washington, D.C., on September 11, 2001, the perfect event for the implementation of fascism. Since I have neither the time nor the space to elaborate, I will speak plainly about this event. Whoever carried out this horrible tragedy, it was not just the accused Arab terrorists, who were first identified because one of their passports was found in good condition on the street below (never mind the fact that the planes they were hijacking were incinerated in the buildings). Without the inside green light from military and secret agencies leading all the way to the White House, it is impossible for groups of terrorists to get through American airports and hijack four planes and attempt to crash them into three domestic targets (the World Trade Center, the Pentagon, and the White House). Period. In light of what happened to the Minnesota FBI agents who arrested Zacarias Moussaoui for acting suspiciously in a Minnesota flight school, this conspiracy is at least partly based in the United States.

In early August of 2001, local FBI agents arrested Moussaoui and immediately petitioned the national FBI for a warrant to get into his computer. Their request to search his computer was mysteriously denied. As we know from Moussaoui's trial in 2006, 9/11 could probably have been prevented if the FBI had been able to examine Moussaoui's computer, since many of the other accused attackers also acted suspiciously in U.S. flight schools. This story about the local FBI agents and the flight school instructors who tipped them off was on CNN constantly during September of 2001 and afterward, so there is no excuse for the public not seeing that three thousand innocent Americans might be alive today if it weren't for the FBI. The Bushite version of who caused 9/11 is full of holes, and readers who doubt this may want to study the huge body of conspiracy literature spawned by the event.[4]

I am forced to focus on 9/11 and the false governmental version of the story because it functions as a critical *timelock* on the whole Galactic Underworld. When I am teaching outside the United States, many students feel that Americans pay way too much attention to 9/11 in light of much greater tragedies, such as the Asian Tsunami of December 26, 2004, and I agree. But the people of the world need to see that because 9/11 stopped time in the United States, the American people are traumatized and unable to rise to the occasion and purge the United States of the Bushites. The intelligence and will of the American people has been seized by 9/11, which splits them into two hopeless factions—those for Bush and those against Bush—that the media trigger into constantly opposing each other.

Meanwhile, the Bushites roll along merrily, running the Galactic acceleration of the whole world. American citizens must face up to the fact that thousands were murdered just so George Junior could beef up the military to invade Afghanistan and Iraq in order to control oil in the Middle East.[5] They certainly aren't running this program to get perfume, so we need to add the crisis of peak oil, which I will follow during the Days of the Galactic.

Regarding the remainder of Day Two, the invasion of Afghanistan to find Osama bin Laden ended the year, and the Patriot Act, supposedly designed to make it easier to catch terrorists, was quickly jammed through Congress so that the destruction of American individual liberty could begin. To express themselves, Americans waved flags, bought diamond, ruby, and sapphire flag pins, and hung yellow ribbons on trees, while foreign bodies were being torn apart by bombs made in the United States.

Unbeknownst to the average American, much of the rest of the world was watching with horror at the outright takeover of America by neocon fascism. Other countries remembered going to war to stop Hitler, and some began to fear that someday they'd have to stop the insidious, bloodthirsty regime in the United States.

Next, we will consider what was going on during Day Two of the National and Planetary Underworlds. To look for parallel developments, I must point out that now we can see that the events of Day Two during the Galactic were the beginning of a great religious war between East and West. Bush even slipped up once in public remarks and used the term "crusade" to define America's plans in the Middle East. As you will see in the Day Two analogs, religion is also a driving force, since it has been the favored tool for manipulating the public for five thousand years. In other words, most Americans can't think straight about 9/11 because they are subconsciously caught in crusading against the evil infidel—the timelock.

Looking back to Day Two of the National, 2326–1932 BC, the story of Abraham and the Patriarchs is the key theme and the basis of Judeo-Christianity. I point out this strain (instead of Egyptology, for example) because my chosen theme here is to trace America as global empire during the Galactic, and America is intensely supportive of Israel as a homeland for Jews. The great father Abraham is the founder of Judaism. He is later adopted by Christianity and then by Islam, so all three religions are *Abrahamitic*. As Calleman notes, the biblical patriarchs living in Day Two "brought the belief in this Creator God from Chaldea to Canaan," which shifted the emphasis out of Sumeria.[6] Looking back to Day Two of the Planetary (AD 1794–1814), we find that once the thirteen states had been established, Americans began moving west (the Louisiana Purchase was in 1803).

The French Revolution and the rise of Napoleon was the big story in Europe then, which was an issue in the United States because many Americans empathized with the rise of the people in France. Many

Americans believed that the French had the courage to overthrow autocracy because the French had seen the Americans do it first. Thomas Jefferson served as president from 1801 to 1809, and overall this was a formative and positive time of growth in the United States. In other words, Day Two was a major phase in the germination of American democracy, and the United States even declared war on Britain in 1812, a war that ended at the close of 1814. Calleman notes that the Napoleonic Wars "shook the established order of the royal houses of Europe in such a way that order could never again be taken for granted."[7] Democracy was in its ascendancy.

NIGHT TWO OF THE GALACTIC UNDERWORLD: DECEMBER 20, 2001–DECEMBER 14, 2002

Night Two of the Galactic was a period of great economic instability caused by the NASDAQ crash, which conveniently drew attention away from the war drums beating in Washington, D.C. People were so traumatized by 9/11 that they longed to go back to the good old days, but that was not possible because they felt threatened financially, and they couldn't believe the country was being drawn into a war. It is interesting that during Night Two of the Planetary (AD 1814–1834), Europe was swept by a wave of Romanticism and the desire to return to the mythical past after Napoleon finally fell in 1815.[8]

The Bushites and the media created a climate of unending fear, and people bought Hummers and SUVs so they'd feel safe in their cars with their cell phones. Calleman notes that during Night Two, "The older order of the economic, military, and media domination of the West was firmly reestablished."[9] Just as during the Vietnam era, anti-American sentiment was building in Europe, while the French struggled to find a way to calm Bush down. Saddam Hussein struggled to prove Iraq did not have weapons of mass destruction (WMD), and Hussein obviously believed the United States knew he'd destroyed his weapons after the Persian Gulf War. We know this because in a CIA secret report released October 6, 2004, it was revealed that the CIA knew there were no such weapons![10] So, regardless of whether the United States had any legitimate reasons for pushing regime change in Iraq, during Night Two the Bushites set up the contingency plans for invading a sovereign nation, while the American public was stunned and fearful as a result of being assaulted by Red, Orange, and Yellow terrorism alerts and anthrax scares.

Depressingly, accounting and corporate fraud of monumental dimensions was being exposed. More and more Americans were losing their pensions, which created growing uncertainty about their futures. Do you remember feeling angry and depressed because a future you worked for was slipping away in an ocean of red ink and bombs?

DAY THREE OF THE GALACTIC UNDERWORLD: DECEMBER 15, 2002–DECEMBER 9, 2003

Day Three of the Galactic—sprouting the seeds—was when the Bushites showed their poker flush, and people just couldn't believe what was in the cards. Exactly when Day Three opened in mid-December of 2002, the war drums began beating furiously to invade Iraq. The United States, against great foreign pressure, recklessly invaded Iraq on the spring equinox of 2003 with a cruel and merciless wave of high-tech bombing the military planners called "Shock and Awe." It slaughtered civilians as well as the world's respect for the United States.

Of course, many Americans were all frothed up by the war being run on television as a fabulous football game. However, many others failed to see why America was attacking Iraq to remove Saddam Hussein, when Osama bin Laden (the supposed mastermind of 9/11) was hiding somewhere in Afghanistan or Pakistan, and most of the hijackers were Saudis. The United States claimed that Hussein was working with al Qaeda, which was never proven, just as WMD were never found. Calleman notes that the Galactic began with an intensification of East–West conflict manifesting as battles between the Judeo-Christian and Muslim religions, and this certainly came out into the open in March 2003![11] There is little to say about this sad and destructive war. By late 2005, a majority of Americans realized they'd been duped in 2003, and that this war and occupation of Iraq would go on for years, draining the strength of the American economy while lining the pockets of a few corporations.

While the world seemed to be going crazy on TV, much more interesting and hopeful movements were going on quietly in America and in the world; the new themes of the Galactic were becoming visible. A revival of fascination with ancient history and the more esoteric aspects of spirituality that had been going on among a small minority began to come to the surface. For example, *The Da Vinci Code* by popular author Dan Brown, which strongly suggests that Jesus was married to Mary Magdalene and had fathered a bloodline, was selling millions of copies.[12] Many books about alternative ancient history were very popular, such as *Fingerprints of the Gods* by Graham Hancock, and millions of Americans were practicing yoga and meditation.[13] Many more people were seeing that the limiting mindsets they'd grown up with were resulting in endless warfare, and they knew they needed to think in new ways, such as by adopting the new historical paradigms. Orthodox lies had compromised their innate intelligence, and possibly the official dumdum versions of the past were the source of humanity's profound aggression. Many people were realizing that *we are much more than we seem to be, and the state of the world demands that we change or we will destroy the planet.* This critical thought was emerging and taking over people's minds.

Climate change emerged as an increasingly big issue during 2003 when the public became concerned about global warming as many areas experienced hot summers and warm winters. There was a terrible heat wave in Europe during the summer of 2003 that killed more than ten thousand people in Paris. To more and more people, especially the French, carrying on wars when the planet was exhibiting such stress was crazy. However, climate issues were not on the neocon table. In light of what America was doing in Iraq, Islamic resistance was organizing and preparing to launch effective resistance to the U.S.-led invasion.

Like Goliath, the United States just sneered at the East, while many a David sacrificed himself as a suicide bomber in Israel and Iraq. Meanwhile, intelligent Americans were getting nervous, and life in America was not so much fun anymore. When people went to the airports, the Transportation Security Administration treated them like cattle going to slaughter; it was obvious that "airline security" was a sham. These incessant fear-based programs aimed at the people made them extremely tense, so they got fat and asked their doctors to prescribe the drugs being advertised daily on television. This was the year of *Super Size Me,* an excellent film about the health risks of eating fast food.

Even the Sun was responding to the pain on Earth; huge solar flares pummeled the magnetosphere in November 2003. Do you remember being hassled by negative energy and finding yourself feeling very angry in 2003? I will look at the analogs during the National and Planetary

Underworlds to better understand the old archetypes we were processing during 2003.

Looking back to Day Three of the National (1538 to 1144 BC), the central theme that sprouted was that of an early monotheism eventually led by Moses. Monotheism is often described as being a great religious advance, but this is a typical interpretation of events by the winners who cooked the books, in this case, the Bible. In fact, monotheism was a belief system that had wreaked war on previous religions consisting of pantheons of gods representing various aspects of the human psyche, as well as making war on the last vestiges of the old Goddess cultures.

One is not better than the other, yet the monotheists were the conquerors who eradicated anyone who didn't agree with them. Now, monotheism is the driving force behind the neocons eradicating the infidel in Afghanistan and Iraq in the name of God. Yet, ironically, the Muslims are also monotheists! During Day Three of the Planetary (AD 1834–1854), the United States was in the middle of a depression—the "Hungry Forties" in the eastern United States—that was triggered by investment panics in 1819 and 1833. A lot of people don't know that America defaulted on its debt to foreigners in 1842, which is mentioned here because the United States is in great debt to foreign countries now.

At that time in our past, new economic engines, such as the railroads and the big bankers, were developing muscle, and the United States was gearing up to play a global economic role with money as its god. The eastern United States dominated the economy of the South, which ended up causing the War Between the States, partly triggered by these great emerging forces. *Religious figures were the patriarchs of the National, industrial moguls were the patriarchs of the Planetary, and IT billionaires are the fathers of the Galactic.*

Do you remember feeling trapped during Day Three by the rising tide of fanaticism that seemed to have nothing to do with what you believed in or cherished?

NIGHT THREE OF THE GALACTIC UNDERWORLD: DECEMBER 10, 2003–DECEMBER 3, 2004

Night Three of the Galactic Underworld was a period of deep unease in America based on a profound loss of direction. The war in Iraq was incredibly stressful, and Osama bin Laden was training more terrorists. While Weapons of Mass Destruction were never found in Iraq, the United States was using WMD on hapless civilians daily. Americans were becoming extremely polarized over being stuck in a quagmire so much like the Vietnam War. Saddam Hussein was pulled out of a rat hole right after Night Three opened, and the American taxpayer would have to pay for his trial. The fundamentalists and lovers of war stood with Bush, since "yer president" is always right, yet for more and more Americans, something was deeply amiss.

The Madrid train bombing on March 11, 2004, carried the East–West conflict into the heart of Spain and showed that terrorists were capable of increasingly sophisticated attacks against civilians. This was a sign terrorism would eventually threaten any country that supported the United States, so Spain withdrew its troops from the "Coalition" fighting in Iraq, making it more obvious that the Iraq War was an American affair.

Then to top it off, sexually sadistic photos, including some of naked male Iraqi prisoners with leashes around their necks being led around like dogs by female American soldiers, humiliated America and inflamed Islam—a religion that goes out of its way to promulgate sexual purity. The euro and Canadian dollar were rising steadily against the dollar, while deficits in the United States were piling up.

Ominously, climate change was beginning to show its destructive power. As if the gods were throwing spears at Bush's brother Jeb, ferocious hurricanes such as Ivan pummeled Florida. Savvy people were extremely worried about what was going on in the Atlantic Ocean. Right at the beginning of Night Three of the Galactic, *Fortune* magazine, an influential source of news for business, ran a long article about the slowdown of the Great Ocean Conveyor, the current in the Atlantic that warms the eastern coast of the United States and the western coast of Europe.[14] As this slowdown is capable of triggering an ice age, people were really feeling the fragility and preciousness of Earth's ecosystems.

The United States just kept on obscenely dropping bombs, and by now it was public knowledge that the neocons were looking forward to the end of the world because they believe that Jesus will return. Since many Christian fundamentalists believe the Jews must rebuild the temple in Jerusalem in the end of time so that Jesus can return, they made alliances with Jewish fundamentalists to rebuild the Temple of Solomon; this alliance lies at the heart of the neocon agenda. For example, Christian churches were raising money to send American Jewish families to Israel to accomplish this end. *Night Three in 2004 was the year when a majority of Americans knew their country was in big trouble and that they were going to suffer the consequences.* After all, the American Constitution is based on the separation of Church and State. This deep unease caused more and more people to investigate alternative versions of the past. The Catholic Church was obviously hoping silence would kill Dan Brown's *The Da Vinci Code,* but more and more, the public just wanted the truth about everything. The genie was out of the bottle, and the historic lies were being challenged everywhere, fueled by the Internet. I was amazed to see three or four people carrying *The Da Vinci Code* onto planes when I took flights during 2004. I mention this because a new understanding of the meaning of the life of Christ was coming in, and this fecund spirituality will build and build during the Days of the Galactic, as you will see. Who says Jesus couldn't have fun too?

Significantly, Carl Johan Calleman's *The Mayan Calendar and the Transformation of Consciousness* was published in the spring of 2004, and knowledge about the mechanics of the evolutionary wave began moving all over the planet like a spiritual tsunami. While the American bully was beating up the world, Calleman's theory was coming forth, offering real hope for the future! As you may remember from my introduction, Calleman's book was sent to me for an endorsement in May 2004, yet I lost my ability to comprehend the implications of it because my son Tom died in June 2004. Tom was carrying *The Mayan Calendar* in his backpack, which was stolen the day he died—the day Venus passed in front of the Sun: the Venus Passage. Calleman's book was put on the shelf while I attempted to comprehend the loss of this second son of mine, and Gerry graciously moved us to Canada so that we could live in a gentler culture. Once we got to Canada, we were amazed to see that large numbers of Americans were buying property in Canada, possibly because they were getting were nervous about the neocon takeover in the States.

Regarding the neocons waiting for the Second Coming, what are they all going to do if Jesus brings Mary Magdalene back with him when he returns? Do you remember during 2004 being angry at the outrageous stupidity and callousness of continuing to kill in Iraq while the planet was suffering an ecological meltdown? Think of what the ramifications may be of pounding Earth with the "Mother of all Bombs." As you will see, during early Day Four, this kind of abuse may trigger tectonic responses.

DAY FOUR OF THE GALACTIC UNDERWORLD: DECEMBER 4, 2004–NOVEMBER 28, 2005

Day Four of the Galactic—the proliferation or spreading of the new themes—was a year of great spiritual awakening. It all began in a huge explosion of human pain, love, and compassion. On Boxing Day, the day after Christmas, a great displacement of the Sumatran fault caused a 9.1 Richter scale earthquake that rattled Indonesia and India and sent a tsunami of epic proportions washing over the shores of Indonesia, Thailand, and India. Because of global media, most of the people on the planet knew about this event as it was unfolding, and they responded en masse. An unprecedented global relief movement was launched in response to this awesome human tragedy, because people's hearts were ripped open by the terrible suffering and the great need of its victims. Anyone who participated in raising money and helping in other ways felt this global wave of love and compassion that united Earth's people.

In Aceh, Indonesia, the United States military was welcomed and praised by people in dire need, and soldiers experienced being appreciated for a change. Meanwhile, the war in Iraq felt obscene and sticky. The world saw the raw and true suffering of the quake and tsunami victims and remembered the fragility of people's lives. Similar suffering was occurring in Iraq, but unlike the tsunami it rarely received media coverage. Public concern about the Iraq War was rising, but the Bushites just kept on going with the media covering the war as if it were a football game; it felt like worlds were splitting. They were! The American public was almost split right down the middle on the issue of being for or against the war in Iraq. If this war had depended on draftees instead of reservists, there would have been big demonstrations against it, but there were few protests because the men and women who were fighting in it had signed up to do so, and young American citizens who opposed the war were afraid of the Bushites.

The United States was losing support and respect all around the world, and in light of Bush's global disfavor, other countries were gaining strength. Suddenly the economies of India and China were thriving, and Chavez of Venezuela was challenging the United States by drawing up alliances with other countries in South America that were becoming radicalized. Argentina had defaulted on its debt to the World Bank with the assistance of a loan from Chavez in November of 2002, and Argentina's economy had boomed as a result. The World Bank and the IMF were weakened, and indigenous leaders were rising in Bolivia and Chile.

A relatively unknown leader in Iran, Mahmoud Ahmadinejad, was elected president in June 2005, and he seems to be very spiritual, as well as very extreme in his views. Because he won during the exact midpoint of Day Four, he may emerge as a world spiritual leader. New members joined the European Union (EU), while Tony Blair led the only European country that had not joined the EU, and he gave significant support to the war in Iraq. The London train bombing was on July 7, 2005—"7/7"—and many people noted the code inference possibly connected to "9/11" and that the Madrid bombing was "3/11."

This leaves little doubt that terrorism with mysterious esoteric roots was intensifying in the world, and everyone was in danger if Western aggression in the Middle East continued. It was obvious to me that the more the West attacked the East, the more terrorism would arise everywhere on Earth; I think anybody who underestimates the esoteric aspects of Islam is a fool.

As if nature had decided to deal with the Bushites, Katrina, a massive category 5 hurricane, barreled into New Orleans, and the levees that keep the city above water broke a few days later. When I saw the terrible scenes of devastation, my heart went out to Katrina's helpless and desperate victims, most of whom were poor black Americans. The world was shocked by the callous lack of response on the part of American authorities. In sharp contrast to the assistance given after the Asian quake and tsunami, the richest country in the world didn't evacuate people effectively in the first place (though Katrina had provided a few days' warning of its coming), and then did basically nothing to help them quickly enough afterward, either in New Orleans or the parts of the Gulf Coast that Katrina also devastated.

This was the big wakeup call for a majority of Americans: With Katrina, many people realized they'd have to take care of themselves instead of believing that the government would take care of them. Some even wondered if the government was happy to see the slums of poor blacks being cleaned out. Katrina was the end of American pride, which was way too overblown anyway ever since the 1960s debacle in Vietnam.

Then Hurricane Rita slammed ashore a month later and added to the suffering. Meanwhile, the United States just kept on bombing away in Iraq.

Oil rose to $75 a barrel after Katrina, and people had to pay over $3 a gallon for gas in the United States. Some smart Americans got rid of their gas hogs and started thinking about how long they could heat their huge homes, yet most people just figured their energy needs would be handled. (Americans are grossly dependent on oil from the Middle Eastern countries that their own country is destabilizing. The United States gets a lot of oil from Venezuela, which is also risky.) China was buying up oil contracts all over the world to fuel its hot economy. The deficit was rising, and more than half of all U.S. Treasury notes were owned by foreigners.

The Bushites just figured they could keep on raising the debt by printing money, since foreign countries were afraid that the U.S. economy would collapse if the dollar crashed. There were many warning signs about this Pollyanna attitude, and all this porcine behavior will come back to haunt the United States within a few years, as you will see.

Amid the hurricanes, quakes, and the tsunami, a deep and quiet spiritual wave was spreading that was capturing the hearts and minds of millions. *The Da Vinci Code* had sold millions of copies and was being turned into a film, and organized religion grumbled about this. Their deafening silence about the book and its new understanding of Christ had not been effective in squelching its popularity, although the clerics could and did refuse to allow filming in some locations. Gnosticism, a more spiritual understanding of Christianity, was flourishing. People were studying the Kabbalah—the Jewish sacred tradition—doing yoga, enjoying Pleiadian Agenda Activations, and contemplating the implications of Calleman's *The Mayan Calendar*.

By May 2005, I awoke from my fog of grief, took Calleman's book off the shelf, and I was stunned. My intention in this lifetime has been to figure out what life on Earth really means, and I have never wavered in my search. Calleman had successfully decoded the evolutionary time wave. I could see that humans could steer the world off its destructive path by intentionally using time waves to change themselves. And the midpoint of the Galactic Underworld was coming in only a month! On June 2, 2005, we would reach the middle of the Galactic, when the spiritual movements that could carry humanity to enlightenment would become visible. There was only one thing to do about this momentous opportunity: Seize it with every ounce of energy I had!

Students were treated to lots of fun in June when I taught a new class called "Galactic Flash." Now that I finally understood what was really happening on the planet and could see how my own work was part of the enlightenment process, I was totally ecstatic when I taught. This was the first time in my life that I had any idea why I was doing what I was doing, and this has profoundly changed my life. Now I can see that we are in the middle of a massive awakening that will sweep away the destructive forces, but only if we seize this building wave of energy.

Back to basics: Remember that wars tend to be created during the Days of the Underworlds, and the war of Day Four was the Bushites against their own citizens! The Patriot Act was in force and being used to harass citizens; Americans were haunted by terrible scenes of suffering old people dying in wheelchairs on freeways in New Orleans; and people were getting sicker and sicker from prescription drugs. The Medicare drug plan, which involved choosing drugs from hundreds of different plans from drug companies, was being foisted on confused and helpless old people. If seniors didn't sign up, their Medicare benefits would be reduced. They were forced to spend many hours with their children trying to figure out how to keep on getting the legal drugs they were addicted to and that the television advertisements made them believe they needed.

This was sad and abusive. This may seem to be just a localized American plan, but this issue is global because of the wide effects of pharmaceuticals, and because people around the world got a chance to see how inept and abusive the American medicine-for-profit system is. I marveled at the cruelty, harshness, and sadness of life for most elderly Americans, and I was actually relieved that my parents were already gone. I also wondered if some old people would dump the drugs as a result of this abuse.

Now that I understood Calleman's theory, I was able to go back into the comparable Day Fours of the National and Planetary to better understand these weird turns of events. Looking back to Day Four of the National (749–355 BC), right at its opening the great Hebrew prophet Isaiah is onstage, warning the people of Israel to change their errant ways and be faithful to their God, whom Isaiah said was also the God of all people on Earth.

The midpoint of Day Four, which is the midpoint of the whole National Underworld, was when some of the greatest spiritual leaders the world has ever known appeared. These were the days of Pythagoras, Lao-tse, Solon, the Buddha, Isaiah II, Mahavira, Confucius, Zoroaster, the early Izapa astronomers, and Plato. *This global spiritual awakening still inspires us now,* and I believe June of 2005 brought forth significant spiritual leaders. As they come forth, time will sort out who they are. For example, Carl Johan Calleman looks to the great teacher Kalki of India, who has been teaching that humanity will attain enlightenment by 2012. Calleman graciously brought my Guatemalan elder, Don Alejandro Oxlaj, to India to see Kalki in early 2006.

Day Four of the Planetary (AD 1873–1893) was when Helena Blavatsky founded the Theosophical Society, Mary Baker Eddy founded the Christian Science Movement, and when the spiritualist movement in America was at its height. Few people know that fully one-third of Americans adopted Spiritualism—a belief in the afterlife and the practice of connecting with spirits—as their religion during the late nineteenth century. All this was very much in reaction to great forces being churned up by the industrial developments that had taken place during the Planetary Underworld.

A lot of people have no idea what a big movement Spiritualism was from 1873 to 1893 because it declined and almost disappeared during the horrific slaughter of the First World War, which caused so many people to lose hope in humanity's future. Corresponding to the Planetary midpoint, during the Galactic belief in past-life therapy, the efficacy of soul retrievals, and the consulting of spirits on the other side became very popular again during Day Four, and so I believe it will grow during the rest of the Galactic Underworld. The film *What the Bleep Do We Know,* which shows the meaning of frequencies and dimensionality, was released during 2004, and it became very popular during 2005. My book *The Alchemy of Nine Dimensions,* which is very similar to *What the Bleep Do We Know,* was published in 2004 and widely read in 2005.[15] Do you remember becoming excited by esoteric and spiritual ideas during 2005 and wondering what your role as a spiritual teacher might be? On a very personal note, my brother Bob Hand and his wife, Diana Hand, opened Wise Awakening in Bellingham, Washington, an incredibly advanced spiritual center that explores healing with sound and frequency technologies. This has become a teaching center for Gerry and me because of Diana and Bob's extraordinary faith in the spiritual awakening of the Galactic Underworld.[16]

NIGHT FOUR OF THE GALACTIC UNDERWORLD: NOVEMBER 29, 2005–NOVEMBER 23, 2006

This book was written during the first half of Night Four of the Galactic Underworld, a time of integration of the great spiritual strides made during Day Four. Everywhere there are quiet yet profound thinkers who are being heard, as they focus on inspiring humanity to embrace oneness, to cease making war on the planet and humanity. The presence of these teachers will become more apparent globally during Day Five of the Galactic, but do not expect to see any of them on television. A great balancing is occurring on Earth that will open human consciousness to enlightenment, which is also causing a great political and economic shift. The Russian, Chinese, and Indian economies are very strong, and a grand unification is occurring in South America. In January 2006, Michelle Bachelet, a female socialist, won the presidential election in Chile. Evo Morales, a popular indigenous leader, won in Bolivia, while Chavez was delivering searing criticisms of George Bush. Bush puts up with the criticisms because he is afraid Hugo Chavez will stop selling oil to the United States.

Incredibly, the Bushites were beating their war drums against another big U.S. oil supplier, Iran. Iran's leader, Mahmoud Ahmadinejad, is becoming a hero because of his open criticism of Bush. Ominously for Americans, Ahmadinejad is an inspired speaker who freely states that the American and British support for Israel is tipping the balance in the Middle East, and that fairness for the Palestinians must come. This, of course, is absolutely true.

The Americans accuse Iran of developing nuclear technology for nuclear weapons, yet Iran insists it is developing nuclear technology for its future energy needs, since it—unlike the United States—is planning for the time when its oilfields will run out. After the debacle over WMD in Iraq, America is being marginalized by Iran. Ahmadinejad taunts Bush in public by saying that the United States can't attack Iran because it is mired in Afghanistan and Iraq, which is the case; America as global empire is losing power. Significantly, during the midpoint of Night Four—May 27, 2006—clear offers to negotiate peacefully with

Iran were being made within the U.N. by France, Germany, the United States, Russia, and China.

In America, support for Bush and his war dipped below 30 percent during the spring of 2006, as the situation in Iraq was turning into a murderous civil war between the Sunnis and the Shiites. Saddam Hussein's trial was an outrageous circus funded by U.S. taxes with Hussein asserting he is still the ruler of Iraq ("regime change" failed), which further fueled the civil war. The world's people grimly watched this terrible pain and suffering, and many people were relieved to see American power diminishing. Major journalists such as Bob Woodruff of ABC News were critically wounded or even dying in Iraq, and I wondered how long journalists could keep their mouths shut about their real feelings. The price of oil and gold skyrocketed in the spring, while the size of U.S. deficits was frightening, which made the world nervous. If the U.S. economy collapsed, the whole world would suffer. Hopefully, the new alliances made possible by the European Union and South American countries working together, plus the development of strong economies in China, India, and Russia, will hold up the world economy if Goliath falls.

What on Earth is really going on?

In early 2006, the writer and mystic Andrew Harvey said at Grace Cathedral in New York, "Humanity is now terminally ill and can only be transfigured by a totally shocking revelation of its shadow side."[17] He believes that humanity needs to realize that the entire world is in the middle of a great crucifixion as species vanish and the patriarchal systems fall. He believes that by understanding the dark night of the soul, humanity can accept the necessity of this crucifixion and trust the "logic of divine transformation"—that is, *trust the time acceleration.*[18] He also believes that those who realize that there is mercy behind the violence will be given extraordinary strength, protection, and revelation and will become spiritual revolutionaries devoting themselves to the preservation of the planet. In my opinion, the reason this transition is so horrific is that the patriarchy must die so the betrayed feminine can again reign on Earth. I equate the feminine with all that is sacred and whole, and I totally trust this process. We must have compassion for our men as the patriarchy dissolves, and we must speak from our hearts about the grave injustices. In May 2006, the blockbuster film *The Da Vinci Code* was finally seen by millions of people in the United States and all over the world. Many people began balancing the male and female aspects of life, yet the film also made many orthodox Catholics very angry. Calleman says that an "irreversible evolution toward wholeness will take place as the Galactic Underworld progresses, and in the process all hierarchies based on dominance—political, religious, or otherwise—will, in one way or another, break down."[19] This is going to get much worse before it gets better, and readers may find using the personal guide for the Galactic Underworld in appendix C to be quite helpful. It is helping me a lot.

Speaking of things getting worse before they get better, unfortunately the great tectonic instability of the Ring of Fire in Indonesia seems to be responding to the great changes. As we approached the midpoint of Night Four, Mount Merapi on Java began erupting. Then on May 27, the New Moon coincided with the midpoint, and there was a great swarm of earthquakes that killed thousands. News commentators noted that this was especially traumatic because Java is now so overpopulated. There were more quakes in this region during July 2006; when the population was smaller, the people could handle it better.

As you will see in the coming chapter, overpopulation is going to be a big issue for the rest of the Galactic Underworld. And the fact that a major tectonic catastrophe occurred in the Ring of Fire during the Night Four midpoint suggests that Calleman's dates are very accurate. It also means that Earth changes are probably accelerating during the Galactic Underworld.

By the midpoint of Night Four, chaos and confusion intensified in America. In the next chapter, in light of some models that may offer insight into this momentous slice of time, I will shift into the future to consider what might be going on during 2007 through 2011. According to Calleman, the Galactic Underworld is the Apocalypse, the time of revelation, which seems to be well underway.[20]

7

ENLIGHTENMENT AND PROPHECY THROUGH 2011

THE GALACTIC UNDERWORLD BREAKDOWN

According to Calleman, during the remainder of the Galactic Underworld—2007 through 2011—all hierarchies based on dominance will break down as the more spiritual lenses of consciousness transmute the narrow scopes of the National and Planetary Underworlds. Now it is time to identify exactly what these hierarchies are, to see how they function at this time, and to look at how and why they came into existence in the first place.

I am the most interested in observing the military, industrial, medical, and religious hierarchies, since they have such a huge impact on the quality of our lives. Hierarchies based on male domination were developed during the National Underworld (3115 BC–AD 2011). Then many people became personally involved in these domination patterns by becoming addicted to the materialistic comforts of the Planetary (AD 1755–2011). By seeing how these dominator patterns influence the collective mind, as well as being honest about our own material addictions, it is possible to imagine how each one of us can adopt ways of life to diminish our participation in these programs. Why bother? *Male domination and materialism is inherently destructive to Earth's ecosystems and the human heart*. We all must consider ending our addictions to material comforts, and consider seeking ways of life in harmony with Earth. As we find new but very old ways of life, intentionally working with the unfolding cycles of the Galactic will prove to be very supportive and enlightening.

If Calleman is right about time acceleration, the Galactic Underworld is a thirteen-year-long reversal of 5,125 years of evolutionary patterns that have caused humans to believe that the solid or material world is all there is. But that is just a *faulty mindset*. Everything that exists comes first from consciousness, since the material world emanates from creative intentions; thus, *consciousness drives evolution*. Reality, as it exists now, is the natural result of how we've been thinking for five thousand years, and *as soon as we change the way we think, material reality will follow.*

As I write, East–West tension is tearing the world apart, yet it all would be different if people just changed the way they think. Actually, things *are* changing, but the catch-22 is that you can't see this ongoing transmutation unless your eyes are wide open to the possibility of miracles. We are recovering the kind of sight we had thousands of years ago, which will enable us to transcend destructive differences between people. The heart is the key to personal survival, and potent electromagnetic waves of the heart are moving into planetary resonance; soon we will be carried away by very intense feelings for Earth.

Calleman comments, "There is hope for humanity, not because we will all suddenly choose to change for the better, but because the consciousness of humanity is subject to a cosmic plan that cannot be manipulated."[1] During the last year, I've been watching people wake up to the fact that there actually *is* an unfolding divine plan that is delineated by the Mayan Calendar. Many people are remembering that humans were chosen to be co-creators with divine intelligence, even if they've never heard of the Mayan Calendar. This is not an arrogant or egotistical understanding of our earthly roles. Instead, seeing that we are meant to play this role demands that we take radical responsibility for our home and our place in the universe. What could that be? To answer this, we each must fall in love with Earth and her creations again; we are meant to be keepers of life, not masters of death.

During the Planetary Underworld, we fell into a very dark left-brain enclosure, and now we must adroitly withdraw our own threads of thought from this limiting mental trap as it collapses. When I grew up in the United States, the majority of the people around me seemed to be completely crazy, and I looked for ways to help people think more clearly. My grandparents carefully trained me to hold the wider indigenous mindset, so I never adopted a lot of the patriarchal way of thinking, and the collapse of this mental trap is making me feel ecstatic. Now I see that more and more people are waking up. During the Galactic, the number of patriarchal schizophrenics and materialistic manic-depressives has been going down, while the number of fresh faces excited about creating their own worlds is steadily going up. I love watching people wake up in the middle of their habitat with magic in their eyes, as if they were new lights on Christmas trees.

The Galactic Underworld is dualistic, yet unlike the National, it favors right-brain perception. This newly opened intuition makes it possible to see the darkness in the structures of the left-brain, dualistic National Underworld. However, we are only in the early stages of breaking these structures down. Since 1999, many people have been unconsciously processing the layer of pain residing in the memories from the 5,125-year-long National Underworld and the 256-year-long Planetary Underworld.

To be absolutely clear, the reason this process is so intense is because of the time-acceleration factor, and because *the National and the Galactic are both dualistic*. Our bodies and minds have only thirteen short years to release the National Underworld's radical dualities—such as East and West polarization—induced into Earth's energy field by the evolutionary pulse. Then, note that Planetary consciousness is *left-brain yet unitized,* thus it superimposed a potent focus of left-brain consciousness right into the National like a polished steel straw. During the Galactic, the intuitive opening means that our "third eye," which sees all dimensions, uses that metallic straw to peer into the true nature of reality. The more rapid Planetary consciousness cuts right through the layers of the National, but it takes the Galactic sight to expose everything contained in the layers.

The systems that developed during the Thirteenth Heaven of the National—1617 through 2011—are especially ingrained in the West, and they are intensely hated by the East, since they were created during a phase of lucrative conquering that legitimized mind-boggling greed. People in the West just don't see how piggy they are and how this angers the East, so the West will have to learn by having its goodies taken away. The development of National civilization culminated during the time when the West successfully dominated the East and conquered the Americas. Now, the West must release this dominance.

Very significantly, European countries near the midline are becoming more balanced because they've suffered enough conquering. As we've seen, America has taken up the role of global empire, and the rest of the world is forced to respond to this reckless American quest. The conquered, such as indigenous people in South America, are drawing together during the Galactic, while the United States is distracted by its blind crusades in the Middle East. The agenda of the Planetary was to develop material comfort, which has been marvelous for some. Now, however, comforts based on oil have become an addiction in the West that can't be supported. In the midst of all this, by the divine plan, the Galactic Underworld is inspiring humanity as a whole to embrace spirituality over materialism.

THE WORLDS OF THE NATIONAL, PLANETARY, AND GALACTIC UNDERWORLDS

There is another cycle in the Mayan Calendar—the Four and Five Worlds—that offers more perspective on transformations during the Galactic Underworld. The belief that we are now in the Fourth or Fifth World is the basis of many Mesoamerican and Native American prophetic traditions, such as the Aztec and the Toltec. Calleman has shed some very interesting new light on the Aztec prophetic traditions by adding his own insights about the Tzolkin and its resonance with the Nine Underworlds.[2]

The Tzolkin can be divided into many subpatterns, such as the Four or Five Worlds, to gain more insight about how light and energy play out through the 260-day lens. Calleman says, "The Fourth World has prepared the ground for the Galactic Underworld, and so at the beginning of the Fourth World, we can discover the embryonic forms of the phenomena that will come to dominate the Galactic Underworld."[3] It turns out that this idea is very significant.

The Four World dates highlight some important turning points in the developmental stages of the Underworlds, especially during the twentieth century. In general, the First World is an initiating phase, the Second World is the foundation phase, the Third World is the most creative phase, and then the harvest is reaped during the Fourth World.

As we already know, the Nine Underworlds describe primary evolutionary developments, and the Four Worlds highlight what is going on in the collective mind of humans, the mysterious zone of archetypes. Since

	National Underworld (3115 BC–AD 2011)	Planetary Underworld (AD 1755–2011)	Galactic Underworld (AD 1999–2011)	Universal Underworld (AD 2011)
First World	3115–1834 BC	AD 1755–1819	Jan. 5, 1999–March 20, 2002	Feb. 11, 2011–April 16, 2011
Second World	1834–552 BC	1819–1883	March 20, 2002–June 2, 2005	April 16, 2011–June 20, 2011
Third World	552 BC–AD 730	1883–1947	June 2, 2005–Aug. 15, 2008	June 20, 2011–Aug. 24, 2011
Fourth World	730–2011	1947–2011	Aug. 15, 2008–Oct. 28, 2011	Aug. 24, 2011–Oct. 28, 2011

Fig. 7.1. *The divisions by Four Worlds of the National, Planetary, Galactic, and Universal Underworlds. (Illustration from Calleman,* The Mayan Calendar and the Transformation of Consciousness.*)*

what people think about creates realities, the influence of these inner mental archetypes is well worth watching. The entry into the Fourth World of the National (AD 730) was during the latter days of the Dark Ages in Europe, when early kings were struggling to rule their territories while coping with the ascendancy of the East. When the Galactic opened in 1999, America-as-empire was involved in conflicts with the East, showing that these old fears of the East were lurking in the collective mind of the West. Then the Galactic time acceleration stirred up these fears in the Western collective mind, and the neocons used them to manipulate the American public into going to war.

During the Third World of the Planetary Underworld (1883–1947), oil afforded awesome levels of material comfort, especially in the United States. Major cities were built in America between 1883 and 1947, when oil was readily available and cheap. Then when the Fourth World opened in 1947, information technologies caused a significant shift in human awareness, and the American suburbs built on cheap oil and gas spread out into the country. Computers were invented between 1946 and 1948, yet notice it wasn't until the opening of the Galactic in 1999 that people could see that computers were the basis of a whole new economy.[4]

The Third World of the Galactic is June 2, 2005, through August 15, 2008, which compares with 552 BC–AD 730 of the National. So, now that you know something about the Planetary and National Third Worlds, imagine the potency of the creative developments of June 2005 through August 2008, the National analog to the time when the world's spiritual traditions spread around the world, and also when many people enjoyed great comfort during the Planetary Underworld.

We need a good example of how Third World archetypes from June 2, 2005, through August 15, 2008, are being processed. Author Dan Brown has been communicating new ideas about Christ in cheap mass-market paperbacks and first-run movies. He has actually accomplished a lot more than that. His earlier novel, *Angels and Demons,* is almost more popular than *The Da Vinci Code.*[5]

I'd like to offer a suggestion on why Brown captivates the Galactic collective mind so much: Because all the deep layers of the National Underworld are being processed, the really big mass events now involve a mixture of religious and political elements. In *Angels and Demons,* Brown successfully describes how religious and political cabals work together in secret to create mass-trauma events. Meanwhile, during the Galactic Third World, the most problematic event is 9/11, because it functions as a timelock during the Galactic. Even though many people sense that, when they try to figure out how the various perpetrators could have pulled it off, they go brain-dead because they can't imagine how 9/11 could have been accomplished. Well, read *Angels and Demons,* and there you will have all the information you need for imagining how 9/11 could have been plotted and carried out. Considering the scope of the Galactic Third World developments, imagine the magnitude of the shifts coming in the collective mind after August 2008, when the harvest of the Fourth World is to be reaped. This I will describe later in this chapter.

Because of the impenetrability of the Dark Ages, we don't know a lot about what was going on in Europe when the Fourth World of the National Underworld commenced. In AD 730, Pope Gregory II excommunicated the Byzantine emperor, and in AD 732, Charles Martel of France stemmed the tide of the Arab advance in the Battle of Tours. Meanwhile, the Arabs were penetrating the West with their culture because the West had lost most of its literature and science during the Dark Ages, while the East had retained significant ancient knowledge. When Eastern consciousness penetrated the educated Western mind, there were constant wars going on between East and West and against the barbarians. For example, the Vikings began sacking Northern Europe around AD 830;

the Arab Caliphate sacked Rome, damaged the Vatican, and destroyed the Vatican fleet in AD 846; and the Mongols appeared in the 1200s out of central Asia. Ordinary life was amazingly dark and dangerous during these times, and *the Western collective mind holds deep hidden fears of invaders from the East and marauding barbarians.* The neocons easily incite the American public to project all these unresolved fears on the East. Meanwhile, the intellectual ferment that has been going on between East and West for thousands of years is what kept cultures and minds alive during the Dark Ages.

This revival of interest in ancient cultures is why the new-paradigm intellectual movement is so important now. Radical new-yet-old ideas are bringing thoughtful Americans out of the Dark Ages; new timelines and paradigms are opening the West to enlightenment. Christianity, for example, is dead without a periodic new view of Christ—a renewed Christology.

Calleman notes that during the Galactic, it is best to think of the countries near the midline—such as Germany, France, Norway, Sweden, and Italy—as a unit in the middle of the intensely dualistic struggle between East and West.[6] Countries close to the midline are often much less polarized; many EU countries opposed America's 2003 invasion of Iraq, for instance. Since the agenda of the Galactic is to unite the world's people and make peace possible, *countries near the midline are slated to be the initiators of peace from now through 2011.* East–West religious wars were a major theme during the National, and the accelerated energy of the Galactic is propelling these old conflicts to the surface to cleanse the old dualities. Initially, this flaring violence caught countries on the midline by surprise, since for most Europeans, wars against the "infidel" are already passé. EU countries are integrating people from the East into their societies, and tensions in the Middle East are making this difficult. For example, the French strongly opposed the American invasion of Iraq because they knew it would destabilize their own Islamic citizens. Europeans can see that the East does not seem to really want to fight. The East is trying to defend itself against the voracious empire that is running out of energy on all levels and wants to rob any country that has oil reserves.

What about the Fourth World of the Planetary (1947–2011), the culmination of materialism? Surely the ultimate goal of material comfort is peace? Remember, the National is about the development of civilization, and the Planetary is about the attainment of human comfort. In 1947, Gandhi's peace movement resulted in India's independence and the partition of India and Pakistan, as well as the independence of major Asian nations such as Burma. Gandhi's peace movement gave hope to the world and was very much admired in the United States; it fueled the resistance to the Vietnam War in the late 1960s. Meanwhile, in light of current East–West tension, the most significant thing that happened in 1947 was the British proposal to divide Palestine. The Jews and Arabs both rejected this proposal, it was referred to the United Nations, and then in 1948 the U.N. founded the state of Israel.

THE MIDDLE EASTERN WHIRLWIND OF VIOLENCE

The formation of the State of Israel instantaneously created a whirlwind of violence amid the Middle Eastern hotbed of religious conflict that was generated during the early Days of the National. For example, Sumer rose during Day One (3115–2721 BC); the Akkadians and the Patriarchs rose during Day Two (2326–1932 BC); the Assyrians and Hebrews under Moses rose during Day Three (1583–1144 BC); and the Babylonians, Persians, and Jews rose during the Fourth Day (749–353 BC). Then Islam, which carries the resonance of the really ancient religions in the Middle East, such as Zoroastrianism from Persia, was formulated during Night Five (AD 434–829). The Middle East is a hotbed of conflict in the collective mind, and *Islam is a blind shadow for Judeo-Christianity* because it arose during Night Five.

So, what is really going on here?

If you read the Quran, you will find it incorporates the Hebrew and Christian scriptures, it has many Gnostic elements, and it adds Muhammad's wisdom to the mix. The Quran is like an evocative spice cake, and actually, you need to hear it sung to appreciate it. Some of the most exquisite mystical moments of my life have occurred to me in Egypt while I listened to prayers from the Quran being chanted. All three of these major world religions—Christianity, Judaism, and Islam—are sourced in very ancient wisdom traditions that first arose in the Middle East during the National Underworld. The land in this area is also soaked with memories and beliefs that have repeatedly caused people to kill each other for God. In 1948, Palestine, which is sacred land for all three religions, was transformed into a global vortex that is processing all the religious issues of the whole National Underworld! [7]

Calleman says peace will finally come in Jerusalem when the whole world has transcended duality, and this will bring a longing for enlightenment. "Every single individual," he says, "has something to do with it."[8] The three major religions are fighting over the same territory while consulting the same scriptures as Abrahamitic religions. This conflict sucks people into a hot collective vortex that incites them to kill others over their beliefs in the same God! Meanwhile, the esoteric elements of Judaism in the Kabbalah are coming forth in popular culture. The esoteric elements of Christianity are emerging in the Gnostic revival via early Christian sources that have been found since 1947, such as the Nag Hammadi scrolls in Egypt. Sufism, the high mystical spirituality within Islam, has influenced many people in the West through Sufi dancing and chanting. *During the Galactic, the awakening of esoteric Judaism, Christianity, and Islam will melt East–West tension and collective fear.*

THE DEATH OF ORGANIZED RELIGION

The whole Galactic Underworld is a time of ending dominance and imbalances, and the three major religions are male-dominant. Because they share the same old patriarchal databank, they reject women's wisdom. As a female spiritual teacher, I have a few suggestions: First of all, religious conflict in the Middle East was mostly generated by European countries outside the region that had previously carved up and divided ancient territories and manipulated internal politics for their own ends. This male way of organizing people by dividing up territories as the spoils of the victor, like Midas sitting at a table counting his gold, has resulted in constant warfare. The fighting between these divided territories has gone on so long that any originating spiritual meaning has long been forgotten, especially since 1755 during the dark material consciousness of the Planetary. Now that the Galactic Underworld has progressed this far, we can see new forms of cultural enlightenment coming from the East, such as yoga from India and delicious and healthy food from many faraway countries in the East; these marvelous cultural gifts are breaking down old divisions. Yet, the West has been eating the food and blowing up the cooks! Eventually, cultural sharing will snuff out the flames of fanaticism.

Second, as we saw in chapter 6, now that America as global empire does not inspire the world anymore, it will lose influence through the rest of the Galactic. The tendency will be for the Middle Eastern countries to control their own issues more, which will be favorable. The idea that powerful countries at a distance can control the destinies of other countries they know little about is a male-dominant idea that is dying in the world. Only Americans are impressed by the word *democracy,* but they've lost democracy at home!

Every person has something to do with the wars and violence on the planet, and the way most people get sucked in is via their personal religious beliefs. During the Galactic, I think that each one of us must take a shortcut and communicate directly with the divine. *We need to remove God from religions lest "he" die in our hearts.* The Pleiadian Agenda Activations access nine dimensions of consciousness—including God—within the students. When we use all nine levels of our minds, contact with the divine is in the eighth dimension (8D), the dimension that receives the time-coded agendas from the Mayan Calendar in the ninth dimension (9D). As we've seen, the 9D Calendar is releasing the plan through the Thirteen Heavens of the Nine Underworlds. These agendas are received in 8D where they can be *apprehended directly* by any person. Meanwhile, as more people understand time acceleration, it becomes very apparent that a pile of stinking beliefs about right and wrong has been fermenting in the collective mind for five thousand years. Like the moment when the fishes came on land or the hominids got up off all fours, we humans are ready to rescue our divine access from organized religions. *All current organized religions are 4D dualistic filter systems that manipulate the human mind for the purposes of the Powers and the Principalities.*

As Calleman has said many times, the Galactic Underworld is the Apocalypse, the time when we will all confront the Beast.[9] Well, the Beast lurks within all organized religions, and his favorite five-star hotel is the Vatican, as is brilliantly portrayed in *Angels and Demons.* These issues are really confusing for people, because religion at the local level can be an exceedingly valuable social support system that creates community. For people who need church, and I believe many do, you must take your churches back, or leave them so they will collapse and you can create your own groups. Once you see how you are hooked into the dualistic collective mind, church doesn't feel so benign anymore. You *can* climb out of the God-box and view the perversions that exist in the high levels of organized religions. Just let your feelings—the agency of your intuition—guide you.

Regardless of what you do during this difficult transition, learn to contact 8D directly and pull your mind out of the hideous violence that is being processed in the 4D collective mind. You have direct access to the divine and need no go-between. Withdrawing your involvement from the old and moldy arguments over the same field of data is what will end the hideous desecration of the human body in the Middle East.

Now, let's combine the themes of chapters 6 and 7 with our knowledge of events from the National and Planetary Underworlds, and the Third and Fourth World influences, to imagine what might be in store from 2007 through 2011. To refresh your mind, the main themes that are being processed during the Galactic are: America as global empire; peak oil; global information systems; medical abuse; war and peace; fundamentalism; climate change; religious control; and new spiritual and intellectual paradigms.

When I transit into the future to imagine how our lives may proceed during the rest of the Galactic Underworld, I incorporate some of Calleman's prophecies, as well as my own; however, no one can predict the future because the world is an ongoing creation. Since so much negativity coming from the National and Planetary is being transmuted during the Galactic, looking at 2007 through 2011 is like having a nightmare. Letting go of old and limiting patterns is always cathartic, however, and opening to entirely new energies is always ecstatic. What follows is totally speculative (please note that this book was written in 2006).

DAY FIVE OF THE GALACTIC UNDERWORLD: NOVEMBER 24, 2006–NOVEMBER 18, 2007

Day Five—the budding phase—is when the *advanced unifying synthesis* arrives and previous Galactic creations actually become visible and begin to take over the field. The issues from the National are creations from AD 40 through 434, when Christianity was formulated and aligned itself with the Roman Empire, becoming a global system that replaced many other beliefs. It is waning now, as the neglected aspects of early Christianity are coming into view. Things will come forth during Day Five that will strip away the power of the Vatican. As we've seen, a new

Christology has been emerging, and many people are fascinated by it. If the Vatican does not quickly embrace the new erotic Christ, belief in the Roman Church will collapse amid the desiccated male hierarchy. Or, the potent fundamentalist movements that are growing within the Catholic Church will take over the structures, and this will marginalize Catholicism because fundamentalists are fanatics.[10] Materialism was god during the Fifth Day of the Planetary (1913–1932). As a result, new expressions of material value will emerge during the Fifth Day of the Galactic. I predict this will be the beginning of the end of growth economies, since we can't grow anymore anyway, yet nations will conspire to get as much oil as possible. Also, during Day Five of the Planetary, we will witness great mass movements, such as the movement of troops during World War I, and pandemics are likely, since the Spanish flu killed millions during 1918–19.

In our time, a great war between East and West may break out, if it has not done so already; this could cause greater movements of people. Unfortunately, as the materialist grip of the Planetary is transmuted, a Galactic pandemic and a great war *are* likely. Mass movements of microbes are often triggered by hubristic human behavior, and ironically, the United States is set up to profit from a pandemic. However, I predict that many people will become much smarter about vaccinations, media health advice, and medical manipulation during the Galactic Fifth Day. Allopathic medicine will be seen as a big killer, and angry people will rise up against it en masse.

As we've seen during the other Days of the Galactic, great Earth changes are likely, and climate change will be a serious issue. Regarding warfare, countries on the midline will emerge as leaders for peace while the United States flounders in a sea of blood and red ink from being overextended in the Middle East. Also during Day Five, the majority of people will realize that the military is the biggest polluter as well as the biggest user of precious energy resources on the planet. People will not be willing to pay for fuel to fly cargo planes and F-16s when they can't heat their own homes. There is no way that Day Five is going to be a cakewalk for anybody—not even for the elite.

While these great transformations of matter are unfolding, great spiritual powers will flood peoples' hearts when they help others survive. The great spiritual potency of the Galactic will inspire people to treasure love and life in the face of so much senseless death. We must learn to face death and dissolution, because nothing new can be born without it. As the loss of life escalates and people feel the great pain of others, they will be forced to face the inevitability of death and appreciate its ecstatic release. My parent's generation, who went through the Depression and World War II, were so afraid of their own deaths that they rarely processed their own emotional problems. When they got old, they were obsessed with avoiding death, the biggest losing battle of all time. For the Greatest Generation, battling death was akin to winning World War II, and so they were taken over by the pharmaceutical companies. The baby boomers have seen the results of their parents' losing battle, and if allopathic medicine continues to be controlled by pharmaceutical companies, the boomers will collapse it by not supporting it.

The most significant revelation in 2007 will be that *the largest avoidable cause of human misery is overpopulation in light of what Earth can provide and what families can handle.* During 2007, a great wave of healthy anger—generated by the blindness of religions to the needs of women and children—will sweep Earth. People will see how religions in general have manipulated human sexuality to prompt women to produce more Christian, Jewish, or Islamic souls. They will also see that religions are the cause of the hideous war that seems to be out of control.

Let's look at population statistics in light of time acceleration. When the Planetary opened in 1755, there were about 1 billion people on the planet. This figure began to rise during the early 1800s, and then once oil was discovered in the 1860s, the population expanded beyond the carrying capacity of the planet, to the current 6 to 7 billion people. During the early Planetary, Thomas Robert Malthus (1766–1834) proposed that unconstrained human population would grow exponentially, and would face strict natural limits.[11] Logically, population planning should have been an essential part of the Planetary search for comfort, but this possibility was globally blocked by the Vatican.

Day Five will be the year when the majority will see that *the population explosion is the ultimate disaster of the industrial world,* and new systems will come forth for balancing the number of people and the habitat. China will be a model for solving this urgent yet simple issue, as well as countries near the midline that are already learning how to live with low birthrates. Eventually, fewer children will mean that each child will be cherished and nurtured by its parents and communities. Adults will be able to access advanced states of spiritual openness from their children—the best reason to have a child, from the parent's point of view. Realizing that excessive birthing is the height of human cruelty will come about when many people see the massive suffering of the poor, who crowded into the world's cities searching for comfort during the Planetary. Now a few billion people are trapped, and their plight will be fully exposed during war and weather disasters. South American leaders, such as Hugo Chavez of Venezuela, will be considered models for how to balance resources between the rich and the poor. The men's Goddess awakening—the realization that women have been used as birthing machines by the patriarchy for thousands of years—will be so profound that people will never allow unplanned births to happen again.

Peak oil—the point when the most easily obtainable half of the world's oil has been extracted—means the world's population will be significantly reduced in a few short years.[12] If this terrifies you, I am a woman who has already lost two of my own children, and I can tell you that you are stronger than you think. My older children are not with me, yet they still exist as souls eternally in my heart. The newly emerging compassionate consciousness of the Galactic Underworld will find kind ways to ease the pain of what is coming so soon. These great losses will mean that only parents who are really qualified in all ways will have children. In a few short generations, all the children in the world will be loved, fed, and protected, and the world will see spirit in their eyes.

The really big and painful wakeup call during Day Five will be in the United States. Many citizens will realize they are stuck in a fascist country that uses their taxes to kill people around the world while their own resources dwindle. Americans will be scandalized when they see how far their government pushes war in the Middle East. The majority of Americans will detest what their government is doing overseas (and on the borders and at home), and they will find ways to withdraw their energy from war and from manufacturing weapons.

During Day Five, many people will see that the American medical system is controlling their bodies for profit—not for health—and they will pull out of the system. Forced vaccinations will be attempted, and this will cause a huge rebellion, like that of draftees during the Vietnam War. The majority will refuse to be injected, and they will go on to question all the other aspects of the medical system. People will just stop taking drugs when they realize that drugs are draining away their long-term health and their money. I am sure this will happen, because some really significant turns in this direction occurred right during the midpoint of Night Four (May 27, 2006), when the key issues of Day Five first emerged. A headline-making study found that Americans are much sicker than the British and the Canadians, even though they spend much more on health care.[13] Conclusion: Excessive drug consumption and excessive medical testing are making Americans sick. I see a massive erosion of the power of the medical mafia during Day Five. Although constant medical testing will go away when people refuse to be laboratory rats, surgical operations and other types of sophisticated medical procedures will remain. Of course, this rejection of the medical mafia will only happen if the people collectively dump the system, and I think they will. If not, America will be divided between those in and out of the medical system, and the more "care" one gets, the sicker one becomes. Also, huge government resources will be dedicated to advanced triage medicine overseas during war, and basic medical care for the American public will suffer.

NIGHT FIVE OF THE GALACTIC UNDERWORLD: NOVEMBER 19, 2007–NOVEMBER 12, 2008

Night Five will be a time of deep integration of all that has gone on during Day Five, which went by so quickly that people could hardly breathe. The changes during Day Five will be the greatest ones of the last five thousand years, because people will begin to think in radical new ways. Then during Night Five, *even though the world will not yet have changed, peoples' minds will have changed*. A huge percentage of the American public will oppose the wars that will be draining America's power. This integration work will be very intense and disturbing, because it involves facing man's inhumanity to man. By now, it will be obvious that powerful and cruel leaders rarely cared about their flocks during the insensitive National and Planetary Underworlds, and the structures from these lower Underworlds will be transformed because nobody believes in them anymore.

The more people just look the cold truth in the eye and begin to change their ways, the better they will be. One good thing I can say about the Galactic is that wars used to last many years, but now because of time acceleration, this process is much faster. The spiritual advances made during Day Five—becoming independent, becoming supportive of others (not systems), and awesome intellectual, emotional, and physical breakthroughs—will cause a minor backlash at first. The media and the controllers will do everything possible to convince people that they are still helpless, they still need drugs, and they can't live without their leaders, but this will not work because the erosion of the dominant systems is ongoing.

The reeling colossus of 5,125 years of patriarchal domination and 256 years of seeking comfort will crash, and during Night Five, this crash is likely to be financial, which also could curb the warfare. Therefore, the way to be independent and have the ability to help yourself and others is to be completely out of debt, own your own home, have many different ways of earning a living, and grow food and develop community. Barter—the real free trade—will become commonplace as governments fall apart because of debt and incompetence, while local governments and communities will strengthen. Calleman says that the processing of National issues during the Fifth Night will "mean a temporary return to the tribe-like world of the nomadic Huns and the marauding Germanic tribes," and that the Night Five Planetary processing will turn up autocrats like Hitler who base their power on superior blood.[14] I think he is right, and I prefer not to imagine how this might work out, but remember this is one short year, not the ten- to thirty-year dark cycles in the past. Whatever transpires, you must remove your energy from the collective mind or you will be swept into the collective insanity, which is most likely to exist in the United States—and this in the middle of the 2008 presidential election! There will be a point when *the old atavistic methods will not dominate because they have been sufficiently processed.* August 15, 2008, is the commencement of the Fourth World of the Galactic, when new possibilities will emerge and totally new creations will come forth.

The National Fourth World (AD 730–2011) saw the heights of divine kingship in Europe and then the eventual demise of this way of ruling. The Planetary Fourth World opening in 1947–48 saw Gandhi's peace movement and the establishment of the Jewish state. During the Fifth Night, it will be obvious to everyone that the Jewish state necessitates a Palestinian state. I predict that, just for sheer survival, the ancient brothers in Israel will make peace between August 15 and November 13, 2008, just because nobody will be able to stomach the violence anymore, and because countries outside of Israel will be so preoccupied with their own problems that they will no longer be involved. Also, as you will see in a moment, it is possible that an outside agency, one that is extraterrestrial, will force Israel and Palestine to make peace.

As I was attempting to imagine the world changing in these ways during early June of 2006, I heard a report on ABC News about a country that has been doing things correctly during the Galactic; this country, Norway, offers a great model that shows that *it is possible to change.* The United States imports vast amounts of oil, and in May 2006, gasoline was about $3 a gallon. Even though it is widely known that we have arrived at peak oil, the development of alternative energy technologies in the United States is minimal. Norway has vast reserves of oil, imports none, and is storing vast amounts. In Norway, gasoline is $7.26 a gallon; looking to the future, the country is developing innovative alternative energy sources. In general, Europe is much more sensible about energy usage than the United States is, and countries near the midline are behaving much more wisely at this time.

So, as the financial crisis comes—and it will—Americans will suffer terribly unless they are smart enough to wean themselves from energy dependency, get out of debt and save, and learn to share with others in dire need. The strength of each family, and I mean all relatives, will matter a lot. Anyone following Calleman's version of the Mayan Calendar has no excuse for not knowing that Night Five is going to be very difficult, and you should prepare for this as much as possible now. Yet, like an exquisitely sung prayer from the Quran, a wave of love will begin to filter in during August 2008, and everyone will respond to it.

Regarding health care, a huge alternative medical movement has been growing in the United States since the 1970s, and natural ways to achieve health are widely known and followed. As the medical and pharmaceutical colossus collapses, alternative health care will thrive. We will take our health back from the medical machine. Of course, allopathic medicine is a huge sector of the American economy, so there will be great economic stress as this occurs. Since alternative medicine is rarely covered by insurance, more and more people will stop paying premiums and pay directly for their care. This will mean less testing and fewer drugs, and people will take care of themselves. Thus, people will be happier and feel better, and this includes the elderly. Our elders are the ones who have been the most victimized by allopathic medicine, and it will be a joy to see them abandon this cruel system and trust their bodies instead.

DAY SIX OF THE GALACTIC UNDERWORLD: NOVEMBER 13, 2008–NOVEMBER 7, 2009

Day Six of the Galactic—Flowering—is when we experience the *renaissance of the advanced unifying synthesis* that first emerged during Day Five.[15] If you are an optimist, as I am, some really exciting possibilities will come forth during this time. To begin with, the last two Days and the last Night of the Galactic occur during the Fourth World (August 15, 2008–October 28, 2011), the harvest of the collective mind. I think this suggests that humanity won't have a collective mind on October 28, 2011, because we will be in an enlightened state.

Regarding Day Six at this point, most people will have accepted the reality that peak oil has been reached, and many countries and businesses will have adopted radical conservation of the remaining fossil fuels. Wasting resources, especially in war, will be considered obscene. People will be using what is left to develop alternative energy technologies that work *with* the powers of Earth. European countries on the midline will be the leaders of the planet, and the euro will be the world's currency, because these countries will have successfully invested heavily in alternative health and energy technologies. In Europe, leaders began planning for the new world because they had studied the Mayan Calendar.

During Day Six, Americans will no longer be consuming most of the world's resources, and investors will be supporting new technologies. The sale of U.S. weapons to other countries will decline, and the United States will not have the energy to continue to deploy its own weapons across the globe. As a result, the armament industry will collapse like a drunken King Kong. Organized religions will have little influence in the world, since many people will have embraced direct contact with the divine and will be involved in conscious co-creation. Leaders of organized religions will have realized that the world is not ending as they always thought it would. They will be waking up and realizing they will have to do things they don't like, such as acknowledging women in order to include them, and building healthy communities.

Natural medicine will be used for almost all health needs, and allopathic medicine will be reduced to only critical care and surgery. The drug era will be remembered with horror, just as the era of using leeches was, and millions of young people will be working as healers for animals, plants, and people. Peace will have come to Israel, and all three religions will be joyfully sharing the sacred sites. *People won't remember what they were fighting about.* There will be many less people on the planet because of the wars, pandemics, and Earth changes, and each elder, adult, and child will be cherished just because they are alive and participating. The Earth changes will still be very intense and difficult, but because there will be fewer people who live in simpler homes, suffering will decrease. Economies will be recovering because the investments in new ways of balanced life will be beginning to pay off. In case this sounds too positive, all of these things will not have occurred yet, but we will be heading in these directions.

As much as possible, scientists will cease taking energy from Earth because they will be able to actually feel the damage from such activities; they will discover that overusing Earth is what causes the storms, quakes, and pandemics. Calleman says that the dreaded Earth changes are the consequence of the old values ending as a result of changes in consciousness.[16] Being with the flow of Earth's expression instead of taking what one wants will be a very difficult transition. Scientists will feel terrible dark energy in their own bodies when they try to misuse the powers of Earth, and they will change after watching their colleagues die. Many will accept this venerable truth, and they will happily let go of the old values as soon as possible rather than suffer more terrible storms and earthquakes. For example, greedy people will finally realize that we can't keep on burning oil in the atmosphere because this makes temperatures unbearable.

Humans will have been humbled, yet they also will be flowing ecstatically with the divine. Many leaders will be using the Mayan Calendar to direct cultures, and many people will be involved in ceremonies and creations to prepare for the Universal Underworld, the coming enlightenment of Earth. How, you say, can I imagine such things? It is so easy because of another critical factor—*exopolitics.*

THE TRUTH ABOUT EXOPOLITICS DURING DAY SIX OF THE GALACTIC UNDERWORLD

The renaissance of the advanced unifying synthesis could happen by means of *exopolitics,* the political system that governs the universe. According to Alfred Lambremont Webre, who is a futurist at the Stanford Research Institute and lives in Canada, Earth is an isolated planet in the midst of an evolving and highly organized interplanetary, intergalactic, and multidimensional universe.[17] Earth is a member of a collective universe that functions by universal law, and life was planted and cultivated here under the stewardship of more advanced societies. If a planet endangers the collective whole, as Earth is so obviously doing, the "Universe government" can remove that planet from open circulation within Universe society.[18] As Webre puts it, "Earth has suffered for eons as an exopolitical outcast among the community of Universe civilizations."[19] That is, Earth was *quarantined!*

I believe that Earth was quarantined when humans became a traumatized species during the 9500 BC cataclysm described in chapter 3. Also, because our axis tilted and most other planets were affected as well, the geometry of the whole solar system must have been knocked out of balance. I think that in 2011 humans will have evolved out of the mass-trauma mindset, and our quarantine will be lifted if we have each prepared ourselves properly. This scenario is not a *deus ex machina,* because it will only happen if we humans take responsibility for our habitat. *Responsibility* means "the ability to respond," and we will be ready because the Earth changes will force all humans to feel the Earth—to respond to her. I remind you of the indigenous people of the Andaman Islands off Sumatra who knew exactly what to do when the December 26, 2004, quake and tsunami swept their home. Nobody is going to fix our species; we have to do it ourselves. That is the whole lesson of the Galactic Underworld.

I turn to exopolitics here because it resonates deeply with my Pleiadian knowledge about what happened so recently to Earth. Calleman notes that the first real wave of reports about UFOs occurred in 1947, exactly when the Fourth World of the Planetary commenced.[20] He feels that this does not necessarily mean such reports are real visits from other planets, and, as you will see in a moment, I don't think so either. But we both think these reports meant that people began imagining that intelligent beings on other planets exist and that we can communicate with them. Using the various turning points in the Mayan Calendar in light of Calleman's theory—especially 1947—major revelations have come forth during these turning points.

I have publicly claimed for more than twenty years that my consciousness exists simultaneously on Alcyone, the central star of the Pleiades, and on Earth. I maintain this attunement by means of multidimensional consciousness, which anybody can do. Readers will have to consult my other books for details about this, but this viewpoint is very relevant to exopolitics, which I have no doubt will be the next stage of global politics, since it is already being taken very seriously by many scientists and people in high positions in governments in the world. Exobiology is a new field in science that occupies a boundary position between astrophysics, biology, engineering, and even sociology.[21] As far as I can tell, the global elite know all about Earth's quarantine, which is what enabled them to control humans in the first place. As you will see, they *like* the quarantine, and they will keep humanity in ignorance as long as possible. First, we must understand what exopolitics is, and then I will add the Mayan Calendar acceleration to this radical new idea. If both these theories are accurate, they ought to complement each other. Meanwhile, I ask you, haven't you ever felt that perhaps Earth is quarantined in the universe?

Webre says that the quarantine has begun lifting since 1947, which is why I predict that the Palestinian–Israeli conflict will be resolved by extraterrestrial influence in August of 2008—the Galactic analog to the Planetary Fourth World entrance in 1947. Webre says that these high-quality UFO encounters in 1947 are evidence of a "leaky embargo," which suggests these encounters are actually intentional "leaks" in Earth's total quarantine.[22] The various Pleiadian channels, such as Barbara Marciniak and myself, are also signs of leaks. Considering how difficult it is to establish any quarantine, imagine what it would take to quarantine Earth from a populated universe. Webre notes this difficulty and comments that Universe government would have to enforce "an interplanetary and *inter-dimensional quarantine* of Earth by applying advanced principles in parascience [italics mine]."[23]

The interplanetary quarantine would involve sophisticated monitoring and blocking technology that we only now can begin to imagine because we are using technologies based on frequencies. Regarding the interdimensional block, it is exactly what we attempt to penetrate during Pleiadian Agenda Activations. This is more or less successful according to the openness of the students. Regarding the quarantine, I have no doubt that we humans are being monitored by Universe society, and the part of my consciousness in Alcyone thinks we deserve it. I am a pacifist, and the human level of hideous violence and cruelty will never be allowed to go out into the Pleiades or anywhere in Universe society. Human violence will keep us blocked from Universe society until we change. As more of us who are kind and gentle seek a loving relationship with the greater whole, this new positive energy is breaking down interdimensional barriers.

Because I am a teacher in this field, nobody knows better than I why this breakthrough has to take time. We have to be careful not to go too fast with students, thus we have one healer for every ten students in our activations. From my perspective in mid-2006, very soon—especially as Day Five opens on November 24, 2006—this whole process is really going to speed up. It has to, because according to the calendar, we do not have much Earth time left. We will either achieve entrance into the Universe society in 2011, or what we know to be humanity will cease to exist. According to Webre, we are being held back now because the Universe society does not want Earth to export war and violence into interstellar or interdimensional space. As he says, "The *militarization of outer space* may be the single most important factor preventing the end of Earth's isolation from civilized space society [italics mine]."[24]

I would like to add why I feel we are being held back: Humans are afflicted with *misplaced concreteness*—the tendency to make nonphysical beings material—because materialistic science eliminated consciousness as the motivating factor in evolution during the Planetary. Needing to have extraterrestrial beings be solid to believe in them, such as seeing UFOs or the Virgin Mary, has eliminated contact with the vast majority of interdimensional intelligence. Extraterrestrials *can* appear in solid form in Earth's dimension, but this forces them to use huge amounts of energy, which is thought of as wasteful by Universe society. During activations, we teach that it is easier to get out of our quarantine by learning how to access and communicate with beings in other dimensions according to *their* frequencies, which results in spiritual ecstasy for humans.

Introducing exopolitics as the renaissance of the advanced unifying synthesis of the Calendar during Day Six of the Galactic means I will not go on to speculate about what will happen during Night Six and Day Seven, since the direction will be obvious when I explain a little bit more about the implications of exopolitics and the Mayan Calendar. Day Seven of the Galactic (November 3, 2010–October 28, 2011) is also when the 260-day Universal Underworld occurs, the acceleration of the Ninth and last Underworld. Since we are talking about the Universal Underworld as well as Universe society, it is obvious what needs to happen during Day Seven of the Galactic and the whole Universal Underworld. We will experience incredible levels of growth, stress, and integration during Night Six (November 8, 2009–November 2, 2010). Watch out for the global elite's misplaced concreteness when they try to frighten humanity by using laser-hologram technology and Hollywood special effects to project images of mass extraterrestrial landings in the deserts. Remember, Universe society would never waste energy like that, so greet it as a very artistic Hollywood stunt and have a good laugh!

The next chapter covers the scientific aspects of what will be going on in our solar system and the universe during the end of the Calendar, as well as the great leap in spiritual openness that is going to come. Here, I just want to comment on the fact that the big struggle for humanity—which will transpire from August 15, 2008, until the end of the Calendar (once peace is made in Israel)—is going to be over the militarization of space. From my perspective as a female spiritual teacher, it is obvious that the evil-minded scientists and world leaders are militarizing space to force the continuance of Earth's quarantine so they can keep their power and control. Since the timing of the Mayan Calendar acceleration is so obviously in synchronicity with the plans of Universe society, I believe *Earth's Sun will erupt to stop the militarization of space.* Massive solar flares and eruptions will occur that will destroy the technology that otherwise could take war and violence out into space. There is simply no doubt about this, and the Earth changes, consciousness changes, and weather changes that the Sun will cause will be truly cataclysmic during 2008 through 2010.

Finally, Earth will join Universal society in 2011 during the Universal Underworld. I am not saying these things to frighten you, but to offer a realistic explanation for what may be going on. I end this chapter by sharing what I know of Earth's quarantine from the Pleiadian perspective. This highlights exactly what each one of us must do to change ourselves now.

A MESSAGE FROM THE PLEIADES ABOUT ENDING EARTH'S QUARANTINE IN THE UNIVERSE IN AD 2011

According to the information I channeled from Alcyone in 1994, the Pleiadians were deeply involved with life and evolution on Earth until the cataclysm in 9500 BC. Anthropologist Richard Rudgely notes that a tradition of communicable knowledge about the Pleiades—actual observation of the Pleiades and its descriptive name, the Seven Sisters—goes back at least forty thousand years to a time when humans developed symbolic consciousness.[25] So the sense is that *during the Regional Underworld in Paleolithic culture, we were part of Universe society.*

I have traveled back into the Regional by using Paleolithic postures, and we did not seem to be cut off from cosmic space then. Webre notes that the Universe society has sophisticated life-technologies—such as determining when planets have the right elements to evolve life and carrying out implantation programs that are used on suitable planets like Earth—and this process takes billions of years. He comments, "As our human scientists uncover more of the scientific past of our planet, they are actually discovering the work products of highly advanced agents within our Universe."[26] So again, why was Earth quarantined? Webre says, "We humans are the children of a universal cataclysmic event suffered in the course of our planetary evolution. Our subsequent planetary isolation accounts for the severely conflicted, violent, ignorant, and confused state of our history and our society. It is no accident that humans are cursed with war, violence, poverty, ignorance, and death. The violence of the twentieth century would not have occurred on a normal life-bearing planet that had not experienced such an evolutionary disaster."[27] I agree, and exactly that cataclysm and the aftereffects have already been described in chapter 3. What the new timeline in figure 3.2 actually describes is Earth's quarantine in 9500 BC, which triggered a human regression until very recently. Webre comments, "Our planetary quarantine allows the full effects of the planetary cataclysm to play out."[28] I believe this has happened and is ending. "Control over evolution," Webre continues, "is activated when the very existence of a desired evolutionary end is threatened."[29] I think the possibility that Earth's desired evolutionary end is threatened is why the World Tree exists and causes time acceleration, and I think the Mayan Calendar may be an ancient exopolitical calendar that goes all the way back to

9500 BC, right back to the early Vedas. The cataclysm and its aftereffects caused humankind to become a multitraumatized species that has been challenged to the max to survive—and we have, and we will. I think the veil is already lifting, and the curtain is going to go up during the Fifth Day of the Galactic. I have no doubt that we will make it. Webre believes that *Earth's isolation has caused us to develop a soul that can survive on hope alone.*[30] I have great hope in the future, and I am an optimist in spite of knowing what our species must go through to attain reentry into Universe society in 2011. Possibly, after all that we have experienced, Earth will be the teacher in Universe society for how to heal from cataclysm and quarantine.

Furthermore, I believe that our Sun responds to Earth's atmosphere, since Earth responds to the conditions on the Sun. We became a violent and warlike species as a result of very recent cataclysms in the solar system. Now, with the militarization of space, the global elite threatens the solar system itself, possibly the Galaxy itself. Because the United States and its allies project their own fears on others and then attack them just because they exist and have something that is wanted, *a greater power in this solar system will restrain this destructive force.* As Webre says, when a desired evolutionary end is threatened, control over evolution is activated. Thus it doesn't make sense that a rich cabal of needy fools could destroy our planet. Our greater consciousness, the Mayan Calendar, demands we remove the quarantine!

As you will see during the next few years, the king of the solar system—the Sun—is in charge, since this crisis involves the whole solar system. But, speaking of being a species that has existed on hope alone, how will we handle the changes in the Sun when the solar lord essentially fries Earth's frequency-technologies? The answer is by developing the real spiritual potential of humankind. Our true spiritual nature links us to the cosmos, which is the real source of our personal ability to hope for Earth's future.

Next, we will consider the evidence for the influence of spiritual guidance of evolutionary processes on Earth.

8

CHRIST AND THE COSMOS

QUETZALCOATL AND THE NINE DIMENSIONS

The implications of Carl Johan Calleman's theory of time acceleration as the driving force of evolution in the universe are mind-boggling, especially since the whole process culminates in 2011 and balances during 2012, which is explained in detail in appendix B. Finally, here are some thoughts about relationships between Calleman's evolutionary theory and some extremely radical ideas about the universe and consciousness.

This final chapter is wildly speculative and complex because it wonders if the incarnation of Christ, angelic intervention in human affairs, and some little-known cosmological entities may be orchestrating time acceleration and human evolution. In other words, it explores the possibility that *spiritual forces that work in the material realm are guiding evolution.* Such an idea requires an explanation of how I came to it myself, since everything I've understood about time acceleration comes from what I've experienced as a teacher and a shaman.

When I first fully realized the global importance of Calleman's theories in June of 2005, I began teaching his work along with my own wherever I went. However, it is unusual for a writer and teacher of more than twenty-five years to suddenly begin extensively teaching someone else's ideas. This has happened as a result of what I'd discovered by writing *The Pleiadian Agenda: A New Cosmology for the Age of Light*. That book describes nine dimensions of consciousness that will be fully open within all living humans by 2012, as well as a tenth dimension that is the corridor of energy from the iron-core crystal in the center of Earth (1D) into the black hole in the center of the Milky Way Galaxy (9D).

Each of the nine dimensions has a Keeper—a form of consciousness that keeps it in form—and each dimension has a spatial location, such as in the Pleiades, Sirius, or the Orion star systems. When this cosmic databank came through me in 1995, I only understood the five lower dimensions. In the beginning, the upper four dimensions—creation by geometry and morphic resonance (6D), creation by sound (7D), creation by light (8D), and creation by time (9D)—made little sense to me. However, whenever I taught this dimensional model, people *resonated* with it, which astonished me. What do I mean by resonate? People could feel in their bodies that this dimensional model was accurate and important. It matched something they already knew inside, and they sensed that understanding it was the next step in their personal awakening. Well, the same resonance occurred when I taught Calleman's ideas during 2005 and 2006. Most students resonated powerfully with his time-acceleration model; they could feel its critical importance as I did. They were struggling with the rapid changes in their own lives, and they felt comforted by being offered a reason for those changes. You may feel some relief from time acceleration just from reading this book, and the "Guide to the Galactic Underworld" in appendix C may be even more helpful.

People resonated so intensely with *The Pleiadian Agenda's* dimensional model that I spent eight years seeking its scientific underpinnings. These findings were published in *Alchemy of Nine Dimensions* in 2004, which is a scientific analysis of a channeled book! Between 1995 and 2003, most of the dimensional theories in the book were verified by new scientific discoveries, such as the existence of the black hole in the center of the Milky Way Galaxy in 2003. All of these verifications are documented in *Alchemy*.

Thus, by 2004, I had finally grasped the meanings of 6D, 7D, and 8D, yet I still could not comprehend the full implications of 9D. What I *did* understand about 9D is that it is located in the black hole in the center of the Milky Way Galaxy, where time agendas are released that generate the processes of creation on Earth—*the Nine Underworlds!* This link between evolutionary acceleration and 9D is why I adopted Calleman's model. This tie-in with the Mayan Calendar took me farther with my own dimensional model, since the Pleiadians say 9D is Tzolkin. As we know, the 260-day calendar is the Tzolkin, so it was a mystery to me why the Pleiadians said this, but this idea at least cued me into realizing that time is located in 9D.

What I did not understand about 9D is why its Keepers are the Enochians. Well, these Keepers may reveal the actual mechanics of evolutionary processes, as you will see. This ultimately led me to the revelation that Christ—Quetzalcoatl to the Maya—is the central motivating figure of the evolutionary processes, the nexus point between the material and spiritual realms. This is not a new idea, since the early Church struggled with defining Christ's human and spiritual nature. Yet I do not think the early Church orthodoxy was able to define this physical/spiritual nexus, probably because political agendas within the Church distorted its view of Christ. Our understanding of Christ is still evolving, and I've already traced the cultural impact of Dan Brown's Christology during the Galactic Underworld. A spiritual breakthrough is right in front of our eyes, yet few can see this while all the unprocessed issues from all the Underworlds are culminating. Well, we are all spiritual as well as material, which is the revelation of the Incarnation. Consider this: If we were merely material, then the driving force of the evolutionary process would be some kind of mechanical time lord, the most complicated and controlling machine ever known to humankind. Meanwhile, if we can identify the spiritual force behind evolution, then I believe humans will instantly become responsible Keepers of 3D. After all, humans were responsible for their habitat during the last one hundred thousand years; it is only recently that we have gone out of balance. Of course, we are different from the Paleolithic people who lacked ego, a necessary phase of developing consciousness.

We are about to take responsibility for being the species that is at the top of the evolutionary chain. When we realize that Christ reveals our true potential, how could we destroy the planet where he incarnated? Even though many writers have had many things to say about Christ, I've only found a few that are saying anything new. To grapple with the meaning of the Incarnation, first we will begin way out in the cosmos, since Christ surely came from such realms. Lastly, we will find our way back to Earth to investigate the real truth about the life of Christ.

GALACTIC SUPERWAVES AND THE 9500 BC CATACLYSM

Paul A. LaViolette is a physicist and cosmologist and the author of *Earth Under Fire,* which investigates the great catastrophe described in chapter 3.[1] LaViolette says a "great *superwave*" came to Earth from the Galactic Center around 14,000 to 11,500 years ago. It altered Earth's climate, and caused the huge Earth changes and the waves of death known as the Pleistocene Extinctions. According to LaViolette, a superwave is an expanding shell of radiation "that travels radially outward from the Galactic nucleus at close to the speed of light" and penetrates through the Galaxy and beyond the spiral arm disk.[2]

Superwaves can trigger stars to become supernovas or hurl comets at stars and planets, which can create catastrophes. LaViolette's superwave theory has caught the attention of some Mayan Calendar end-date researchers because he thinks another superwave is coming soon. Actually, I don't agree with him about that (although he may be right), but I think he is accurate about the superwave that affected our solar system fourteen thousand to eleven thousand years ago. In *Cataclysm!,* Allan and Delair theorize that it was fragments from the Vela supernova that almost destroyed Earth 11,500 years ago, and they also note that "a surprising number [of supernovas] have exploded unexpectedly near our solar system."[3] Based on *Cataclysm!* and *Earth Under Fire,* I think a superwave triggered the Vela supernova event (dated 10,000 to 12,000 years ago, and some say 14,300 years ago), and then the Vela fragments pummeled our solar system, which is only around 800 light-years away from the Vela system.[4] We know from the great climate changes of 14,000 to 11,000 years ago that Earth was impacted by something from space, and the superwave theory explains a lot about what was going on during those few thousand years.

LaViolette does not single out the 9500 BC cataclysm. However, it inspired me to name a syndrome—*catastrophobia*—to emphasize that we became a multitraumatized species and degenerated after the event. The idea is that many people fear a catastrophe in the near future because their minds are filled with visions of the past cataclysm, yet they can't identify these memory contents because science did not describe these horrific events until very recently. I contend that these past memories flash in people's brains (especially when they get caught in bombings and war), and they think they are seeing visions of a soon-to-come future—possibly the most intense process being transmuted during the Galactic Underworld.

Fundamentalists are especially polluted with this mental detritus, which causes them to long for a quick end to life on Earth—the Rapture, the Apocalypse. These unresolved fears that blind the human mind are polluting interpretations of the Mayan Calendar end-date. I don't think another superwave is coming soon because thinking this way is what I call *primary catastrophobia*. I also believe that Earth's axis tilted only 11,500 years ago, and Earth and the whole solar system are in the process of new alignments. There is evidence for this idea, as you can see in appendix A, and many astrophysicists are reporting on phenomenal changes in many of the planets that may be strong signs that the solar system is settling down.[5]

Meanwhile, another area of LaViolette's research—pulsar theory—is exceedingly relevant during the last years of the Mayan Calendar. I had to describe his superwave theory first because he contends pulsars are an extraterrestrial information (ETI) signaling system that is informing us about superwaves in the past and possibly warning us about ones in the future. Pulsars are stars that pulse out radio waves and flashes of light at varying frequencies and intervals; they were first noticed in 1967. Now we've arrived at the moment when it is possible to imagine how ETI *could* signal Earth, because we have technologies—such as particle-beam accelerators and masers—that could accomplish this.[6] Assuming it is true that we are quarantined because we are a multitraumatized and violent species, the existence of a pulsar ETI system adds a whole new level of significance to Earth's recent cataclysmic period. In other words, *some form of ETI may be signaling us, to clarify what really happened to us so that we can understand our own species!* It is time to realize that we evolved beyond being monkeys and simple hominids for a reason.

UNDERSTANDING REALITY VIA INNER CONSCIOUSNESS

To carry on this discussion, I must clarify why I look at things as I do. From 1982 through 1992, I experienced more than one hundred sessions under hypnosis to explore my so-called past lives. Regardless of what goes on during these kinds of sessions, anybody can go within and time travel into the past or the future. This is because the past, present, and future rule our bodies, not our infinite minds. I published these findings from the sessions in *The Mind Chronicles*.[7]

The first volume—*Eye of the Centaur*—came out in 1986, and it explores various Egyptian, Minoan, Druidic, and Middle Eastern lifetimes. Most of the dates and historical events I reported in the sessions were in pretty close agreement with conventional history and archaeology, yet a few were not. After each session, I researched the historical periods and noticed these differences; however, I published the information exactly as it came through to me. After *Eye of the Centaur* came out, as researchers changed historical dates, often the dates moved closer to what I'd received. The same process went on with the next two books—*Heart of the Christos* in 1989, and *Signet of Atlantis* in 1992. The point is, *I was totally amazed by how much I actually knew inside about Earth's timeline.* I invented a name for this inner device—the inner stellar chronometer. Some might say I was a natural reader of the Akashic Records, the psychic records of all events in time. Whatever was going on, I was building up my personal confidence to interpret the time cycles of Earth.

This process accelerated as a result of receiving *The Pleiadian Agenda,* studying it scientifically, and then writing *Alchemy of Nine Dimensions. Alchemy* tracks this verification process, which I hadn't done with *The Mind Chronicles*. Meanwhile, I wondered how a working mother raising four kids who knew almost nothing about science could channel a book like *The Pleiadian Agenda*. And its wild scientific theories were verified one by one after it was published! To top this all off, when I received *The Pleiadian Agenda,* I was the acquisitions editor at Bear & Company. I discovered while working there that at least four other authors had gotten the same information during the same time frame. I received one finished manuscript, two detailed outlines, and one phone query describing nine dimensions of consciousness—located in the Pleiades, Sirius, Orion, the Andromeda Galaxy, and the Galactic Center—that were coming our way by means of a so-called Photon Band (a band of light particles) from 1987 through 2012! This is not just New Age "love and light," and I still meet people who say they got similar data from 1994 to 1995; cosmic data downloaded into the Earth plane that people resonated with.

I now have a profound level of trust for what each one of us actually knows inside. This chapter discusses things that I know are right because they are accessed from that place of deep intuition. This ability to access information in this way is a potentially valuable approach because there may not be much time left.

What about the time that *is* left? The Mayan Calendar ends in 2011 through 2012, and many people think time will cease in 2012. Coming to the end of time is not about coming to the end of life, yet surely it means *the end of gathering information about our story*. The last things we need to know about history and evolution are very important. As for my thoughts, I think we need to figure out what the Powers and Principalities (global elite) are really planning for Earth. For example, what if the elite know all about an ETI pulsar system that is signaling Earth? What if they are planning to block these pulsar signals with the "Star Wars" missile defense system? What if Star Wars was functioning in 1995, and it blocked the transmissions from the Pleiades? What if Star Wars could block angelic involution, and what if that is what's still going on?

If we can function with nine dimensions of consciousness (as well as activate many other major teachings), I believe we will have our freedom to be readmitted into Universe society, as described by exopolitics. The elite won't be with us (unless they also wake up and open their hearts) because their whole world is built upon blocking human freedom while gobbling the habitat. This is why they (and we) are quarantined.

I think what will happen during the Universal Underworld in 2011 is that the elite control program will end! Since the elite operate in secrecy, attaining certain levels of knowledge is the way out of the frequency jail they've constructed for all but their own. As far as I can see, the big thing they don't want you or me to figure out right now is that pulsars are communicating with Earth. But they have failed in this, for that information is already in print in *Decoding the Message of the Pulsars* by Paul A. LaViolette.[8] More to the point, however, massive solar flares will prevent the militarization of space during 2008–2010.

PAUL A. LAVIOLETTE AND THE MESSAGE OF THE PULSARS

I've been thinking about Paul A. LaViolette's ideas for more than ten years, and his pulsar theory is very relevant for this book. If pulsars are extraterrestrial information (ETI) signaling systems, then this is *the most significant cosmological news of all time,* which must become public knowledge during the Galactic Underworld. This would mean we are not alone; ETI civilizations are signaling Earth to indicate their existence and interest in us.

When the Cambridge astronomer Jocelyn Bell discovered the first pulsars in 1967, he seriously thought he was detecting intelligent extraterrestrial signals, and he named the first pulsar LGM-1 for "little green men." He felt he could not even make an announcement about this, however, without consulting higher authorities, and he even wondered whether or not it was in humankind's best interest just to destroy the evidence.[9] In the early days of investigating pulsars, many astronomers believed the pulsars to be ETI signals. Of course, the government put the lid on such ideas, which was easy; it was made clear that pursuing ETI communications would be the end of one's career in government-funded science. By the mid-1970s, it was anathema to discuss ETI theory; orthodoxy posits that pulsars are spinning neutron stars that resulted from supernova explosions. LaViolette shoots these orthodox theories full of holes and just goes ahead and decodes pulsars as ETI signaling devices.

Of course, you will have to read *Decoding the Message of the Pulsars* for his full hypothesis. In an earlier work, *Genesis of the Cosmos*, LaViolette reveals how esoteric traditions, such as astrology and the tarot, are encoded with clear cosmological, geological, and evolutionary information.[10] He says the Zodiac (constellations on the ecliptic) tells the story of creation and destruction in the universe from Earth's perspective, and that the Zodiac and its accompanying star lore are the remnants of very advanced human preflood (or precataclysmic) knowledge.[11] That is, astrology is *antediluvian* or preflood data.

Astrology uses a 360-degree wheel of 360 Sabian symbols, which is the same as the Divine Year of the Mayan tun-based Calendar and early Vedic cosmology, as already discussed. If the Zodiac is preflood knowledge, then it is of great significance and offers the possibility that the ancient people were very advanced spiritually. LaViolette also contends the Zodiac and the star lore of the Sagittarius constellation indicates that its original designers knew the location of the Galactic Center, which suggests the ancient ones were galactocentric![12] Unlike most astronomers, LaViolette respects astrology for what it is; he is comfortable using the geocentric perspective (things viewed from Earth). Of course, he also knows that Earth orbits the Sun, which orbits the Galaxy. *The geocentric perspective, after all, is what makes it possible to see that pulsars may be signaling Earth.*

PULSARS AS ETI SIGNALING DEVICES

Investigating pulsars as ETI signaling devices, LaViolette concludes that "certain pulsars are nonrandomly placed in the sky, with particularly distinctive beacons being situated at key galactic locations that are meaningful reference points from the standpoint of interstellar communications."[13]

Wondering how galactic civilizations could or would try to reach us, LaViolette began noticing some very unusual cosmic activities. If other intelligent people exist in the universe, then mathematical and geometrical relationships and significant astronomical reference points would be logical ways for them to connect with us. The most obvious one is the Galactic Center, since all civilizations in the Galaxy would consider it a center. For example, as you can see in figure 8.1, the designers of the Pioneer 10 space plaque had similar ideas in mind when they tried to indicate Earth's location to anybody who might find the space probe.

LaViolette saw that the pulsars that were the most noticeable from Earth's view were often close to the one-radian marker.[14] He concluded that this was how an ETI civilization would try to get our attention, and then he noticed some very significant pulsars. The fastest pulsar in the sky—PSR 1937+21—lies closest to the Galaxy's northern one-radian point on the Galactic equator. If you could hear its flashes, it would sound like the musical note E above high C, and its flashings are more precise than the best atomic clocks.[15]

Fast pulsars are called "millisecond pulsars," and PSR 1937+21 is called *the* Millisecond Pulsar because it is the fastest and most luminous, and it emits flashes that are visible with optical telescopes, which is very rare.[16] It is an ideal "marker beacon" because its optically visible pulses facilitate easy detection, and it also regularly produces high-intensity pulses known as giant pulses.[17]

The locations of the star Gamma Sagittae and the Vulpecula pulsar—PSR 1930+22—suggest that the maker of this system "would have had to know how the night sky appears from our particular galactic locale."[18] In other words, *the designer of this system knows about our geocentric perspective!* Another distinctive millisecond pulsar—PSR 1957+20—is

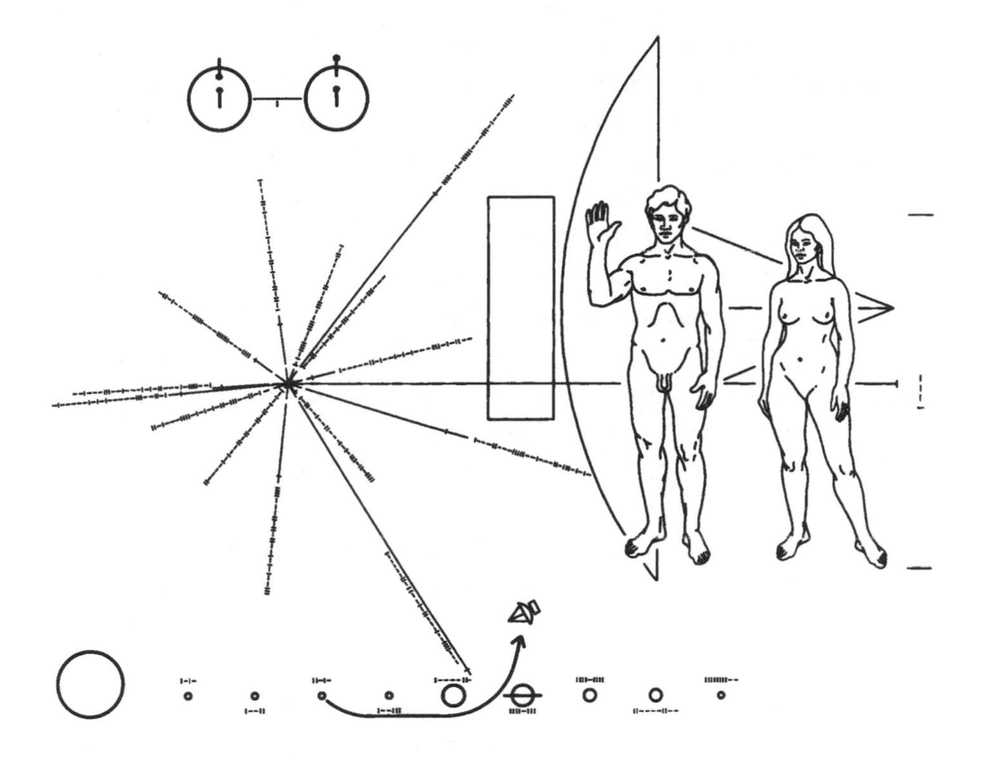

Fig. 8.1. *Pioneer 10 shows Earth's location to extraterrestrials*

located in this key part of the sky in Sagittarius, and its pulsing period is nearly identical to the Millisecond Pulsar.[19] PSR 1957+20 is an eclipsing binary pulsar whose orbital planes are oriented edge-on in our direction, and its companion dwarf star periodically passes in front of (eclipses) it and occludes its signal.[20] PSR 1957+20 also has giant pulses. These very noticeable and unusual entities in the sky really grab our attention on Earth! And LaViolette says, "There is one chance in 10 to the 28th power that this arrangement of pulsars is due to chance."[21]

PULSARS AND EXOPOLITICS

By briefly summarizing such complex insights, I do small justice to Paul A. LaViolette's pulsar theory, and readers must study him to grasp the implications of what he is saying. Furthermore, he wonders if pulsars are signaling equipment on neutron stars that are arranged so they can be viewed from our solar system. His conclusion is, "A galactic civilization of unusually high advancement does exist and is attempting to communicate with us," which would be very significant for exopolitics.[22] The radio pulses now emitted by the Millisecond Pulsar left it 11,700 years ago (because it is 11,700 light-years away), which makes me wonder if it began signaling us just before or during the cataclysm, and we are only seeing these signals now. Are these signals awakening our minds and inspiring us to process repressed cataclysmic trauma? I've suggested that our species will cease being violent and destructive once we identify and process the memories of the catastrophe. Then the quarantine will be lifted and we will be invited back into Universe society.

LaViolette contends that these pulsars signal the progress of a galactic superwave that moved out from the Galactic Center around fourteen thousand years ago through AD 1054, when the wave passed through the Crab Nebula. He speculates that the section of the Zodiac showing Orion holding his shield up to the charging Bull in Taurus may hold warnings about a *future* superwave, as well as an odd optical jet in the Crab Nebula (which is closely parallel to Earth's ecliptic).[23] He speculates, "Could the civilization that engineered the Crab supernova remnant and its unique pulsar beacon also be demonstrating for our benefit a technology that might one day protect us from the onslaught of the next superwave?"[24] Then, assuming we may be being warned about something coming, he offers some ideas on how our scientists could build force-field technologies that could deflect superwave cosmic rays.[25]

Regarding exopolitics, he muses, "If a planetary civilization were to successfully defend itself against such a galactic disaster and survive as a peaceful society without lapsing into a chaotic 'dark age' period, would it then be considered worthy to be removed from quarantine and admitted to the galactic federation?"[26] I submit that there is *minimal* development of peaceful technologies by Earth's governments; in fact, the global elite acts like the charging bull in Taurus! There is every sign that secret governments are doing everything possible to keep us quarantined, and LaViolette even offers evidence for secret military testing of these kinds of technologies. For example, people have seen strange luminous spheres in the night sky—plasmoids—that are probably generated by microwave beams.[27]

Why is all this so secret? Using up the last financial resources of societies for the militarization of space with Star Wars would be the final cataclysm, unless, of course, our governments removed the veil of secrecy and asked for public support for such incredible expenditures for Earth's protection. It seems to me that continuing to go down the Star Wars road would be the final proof to ETI civilizations that Earth's quarantine cannot be lifted. As far as I'm concerned, it's time to stop looking for enemies and process the deep memory contents that cause humans to approach life with fear. Contemplation of this led me back to the strange Watcher angels in the Bible, since the survival period right after the cataclysm is the best place to look for why humans give in so readily to fear. Following the cataclysm, while in a degenerated state, humanity invented religion, which is based on the story of the Watchers and their god, Yahweh.

RECOVERING OUR SPIRITUAL POTENTIAL AND THE WATCHERS

I believe that recovering our spirituality is the only way we can reenter Universe society. I've already suggested that the recovery of the Regional mindset from before the cataclysm—the preflood period—could help us regain our spiritual powers, and so I've been seeking traces of the preflood mentality. These traces lead way back in time to the great and ominous Watcher angels, who are mentioned in the Book of Daniel in the Bible.[28]

In the Biblical lore as well as other sources I describe, the Watcher angels had great bird wings and mated with human females. The Watchers are also deeply associated with the mysterious Enochians—the Keepers of 9D—and the best source for them is the Book of Enoch, an early Jewish sacred text. Often feeling as though I am spending too much time cleaning out a deceased minister's dusty attic, I've kept searching for information about the Watchers and the Enochians. In *From the Ashes of Angels,* Andrew Collins traces the evidence for these bird/human angels all the way back to 8870 BC (right after the cataclysm), to Shanidar Cave in a Kurdish region of Turkey.[29] Shanidar Cave is one of the most important archaeological finds on the planet, with sixteen occupational levels that date back to one hundred thousand years ago; thus it was occupied during the whole Regional Underworld.

The paleontologists Ralph and Rose Solecki found articulated wings of vultures in the 8870 BC layer, which archaic shamans would have used for metamorphosis into vultures during their rituals.[30] They interpreted this finding as evidence for early vulture shamanism, which suggested to Collins that this cache represents the intrusion of the Watchers into a cave that had been a winter habitation place for nomadic tribesmen in the region for ninety thousand years.[31]

As vulture shamans, these Watchers would have been semidivine humans, since becoming a bird in hunter-gatherer cultures confers spirit access. This is still the same today. One night in Kiva East of the Cuyamungue Institute, Felicitas Goodman handed me her cape and asked me to dance the eagle. Circling the sipapu (hole to the center of Earth) and swooping up and down, I became an eagle. I have experienced this potent access to ancient consciousness, which helps me to understand the Watchers.

Eventually, according to the Book of Enoch, the *Nephilim*—the gigantic and degenerated offspring of the Watchers—were born from the daughters of men, and they ran riot and caused great pain among humans.[32] The stories of these giants go way back to the horrific survival period right after the cataclysm, and our minds are filled with unpro-

Fig. 8.2. *The metamorphosis of a vulture shaman.*

cessed bad memories of them. The famous site Jericho in Palestine/Israel dates back to 9000 BC (right after the cataclysm), and the archaeologist Kathleen Kenyon studied six thousand years of habitation there. In the 6832 BC layer, a slaughter of infants was discovered, which the Sumerian scholar Christopher O'Brien suggests may be evidence for a wipeout of the Nephilim, the renegade children of the Watchers, or the murder of children by the Nephilim.[33] Wiped out or not, the Nephilim genes are part of the human race.

Child abuse, murder, and rape are not normal human behavior, and I think these behaviors originated in the horrendous suffering that occurred after the cataclysm. I think *the unprocessed Nephilim within* may possess the minds of the most hideous of criminals. When such archaic contents in the deep mind distress disturbed people, most therapists have no idea what is going on, so they can't reach or help the worst criminals. I believe this could change some day, when therapists have learned more about these archaic shadows. Speaking of Dan Brown's genius for revealing deep, hidden archetypes, Silas the Assassin in the film *The Da Vinci Code* is a classic portrayal of a Nephilim degenerate who was very useful to the Catholic Church; he was a natural killer because he had no understanding of the deep levels of his own archaic mind.

The most complete stories of the Watchers and their offspring, the Nephilim, are in the Book of Enoch, which was a very highly esteemed and treasured Jewish spiritual book that was written down before 200 BC. In AD 325, the early Church father St. Jerome declared the Book of Enoch apocryphal, yet the canon retained most of Genesis (which says very little about the Watchers). Books declared apocryphal were often judged to be heretical, and copies of the Book of Enoch mysteriously disappeared; for more than 1,500 years, this major source for the Watchers was lost. Even the Jews lost this treasured record, which later works occasionally mentioned.[34]

The Scottish Freemason James Bruce found a copy of the Book of Enoch in Axum, Ethiopia, in 1762, which he immediately translated into French and English. More versions of the Book of Enoch were found in 1947 among the Dead Sea Scrolls at Qumran, and now this book is being seriously studied again. It contains extensive information about preflood civilization, and it describes aspects of the cataclysm and the suffering of the Watcher angels; the monster is out of the cage, so to speak. All this is very fascinating, but still I couldn't see how this explained why the Enochians were the Keepers of time and 9D, so like a heat-seeking missile, I just kept on searching.

THE ENOCHIANS AND URIEL'S MACHINE

Next, I read *Uriel's Machine* by the Masonic scholars Christopher Knight and Robert Lomas, which deepened my understanding of the Enochians.[35] Uriel was an archangel (which means Watcher way back in the mists of time), and he is a prominent figure in the Book of Enoch as well as in Masonic rituals. In the "Book of Heavenly Luminaries," a section of the Book of Enoch, Uriel shows the patriarch Enoch a fantastic shining white structure that could observe and measure the movements of the Sun and Venus, which sounds like a sophisticated horizon declinometer. Knight and Lomas believe this shining white structure is Newgrange, a white-quartz megalithic temple near the River Boyne in Ireland, which is close to Tara, where ancient Irish kings were crowned.[36] Newgrange is famous because the rising Sun during the winter solstice illuminates spirals on the back wall of the inner chamber. This temple could be the one that is described in the Book of Enoch, since, as you will see in a moment, there is a lot of evidence for cultural contact between the Middle East and the British Isles five thousand years ago.

Gerry and I visited Newgrange in 1996 right after celebrating the summer solstice and a Pleiadian Agenda Activation at Findhorn in northeast Scotland. We were pretty high after this experience, but that did not explain the energy we both felt in the Newgrange chamber. It felt womblike and safe, and we could feel the presence of very ethereal beings. I could feel my brain downloading ancient data (often my brain responds like a computer when I visit megalithic sites), but none of it made much sense until 2000, when I read *Uriel's Machine.*

Like Newgrange, many megalithic sites were first constructed during the First Day of the National Underworld. Based on an extensive study of Irish and Welsh mythology and clues from Freemasonry, the authors believe this beautiful chamber was used for more than astronomical sightings. Many fragments of ancient beliefs discuss a process for passing the souls of deceased shamans and kings into newborns to provide the community with a flow of wise reincarnated souls, which may have occurred in these kinds of chambers.[37] But what evidence is there for this kind of process?

Freemasonry, which is partly derived from ancient Egypt, associates the light of Venus with resurrection. For example, the kings and pharaohs of ancient Egypt were the sons of God because they were resurrected in the light of Venus rising.[38] The old Welsh *Triads of the Island of Britain* tell the story of Gwydion ap Don, the mother of a sacred tribe who gave birth to the Children of Light, who were astronomers and knew the secrets of agriculture and working metal.[39]

Well, these are the same as the skills of the Watchers that are described in the Book of Enoch, and they are classic preflood skills reported by almost all ancient cultures.[40] Julius Caesar wrote about a Druidic belief in controlled rebirth (passing the soul of the deceased into an infant), and modern Druids say they inherited this belief from their ancient forebears.[41] So what? Well, Knight and Lomas make a strong case that the Book of Enoch and Freemasonry contain evidence for a highly developed astronomy of Venus that was used to pass souls of great leaders into newborns.[42] But why would they bother with this?

PASSING SOULS TO PROTECT THE PERENNIAL WISDOM

Organized soul-passage was probably how ancient cultures retained the accumulated knowledge of their ancestors. This esoteric science goes way beyond reincarnation: It was a technique for manipulating souls to return in new bodies to carry on the wisdom traditions of the past. Sound outrageous? Well, why do we live in a world that is rapidly degenerating even though most people are good-hearted and are trying very hard to live ethical lives? Early Christians believed in reincarnation, and the Bible reports that Jesus was Elisha and John the Baptist was Elijah returned, but reincarnation was declared to be heretical by the Church in AD 553.[43]

The early Church fathers wanted total control over individual salvation, and all this certainly explains why the Book of Enoch was hidden. Regarding my work with past-life regression under hypnosis, my therapist, Gregory Paxson, dedicated his life to regressing people to help them recover their accumulated past-life knowledge. Then they could progress further in this life instead of having to learn things all over again in each lifetime. Wouldn't it be easier to infuse infants with the perennial knowledge at birth? Wouldn't the fastest way to do this be the incarnation of great sages into newborns? I think this is going on right now; great sages are being reborn as the so-called Indigo Children, the return of the Children of Light.[44]

The Dead Sea scrolls were discovered in AD 1947, and they are a major source on the Sons of Light or the Sons of Dawn (Venus), which are Druidic and Essene priestly titles.[45] The scrolls describe two Hebrew priestly lines that existed just before Jesus was born—the Enochians and the Zadokites—but the Romans snuffed out the Enochians during the genocide of the Jews in AD 66, not the Zadokites.[46] Later, the Book of Enoch was removed from the canon as mentioned. Why did the Romans take it from the Jews in the first place? Even more strangely, the Book of James was relegated to the Apocrypha, even though James was the brother of Jesus and took over the early Church after Jesus was crucified.[47] Well, James 21:1–3 says that Jesus was born in a cave to the light of a bright star that went before the wise men, which, of course, is Venus.[48]

Knight and Lomas contend, "Herod feared both Jesus and James, because they were the focus of the rising of an ancient Canaanite cultic tradition which could undermine the Herodian authority if they were allowed to become leaders."[49] Of course, this ancient cultic tradition must have been the Venus birthing rituals, which must have been used to engineer soul involution for Jesus!

Lomas and Knight found Masonic rituals that "introduce *Jesus as the principal character in place of Enoch* [italics mine]."[50] Well, what if Enoch's soul was passed into Jesus? That would be a good reason for the Romans to take the Book of Enoch away from the Jews, since without it, they would not recognize the Messiah. Knight and Lomas wonder if the passage in James means this was the same ritual they believe was performed in Newgrange three thousand years before to pass the soul of the king into a newborn.[51] For this to be true, there would have to be evidence for early Canaanite contact with the British Isles five thousand years ago, and as you will see, there is. This potent Canaanite kingship rite certainly explains why Herod was so afraid of this infant adored by the Three Wise Men (the three astrologers), and why Herod massacred the first-born of Israel to snuff out this line. This infant would have power greater than Herod's or Rome's. According to the new-paradigm writer Gordon Strachan, there may be even more to it than that.[52] I leave the topic of *Uriel's Machine* by mentioning that Knight and Lomas discovered that Venus during the winter solstice (when the solar light comes into the Newgrange chamber and into many other megalithic temples) was in the same position in AD 2001 as it was in 7 BC, when Jesus was probably born.[53]

JESUS CHRIST AS PANTOCRATOR

In *Jesus the Master Builder,* Strachan uncovers a significant network of ancient connections between the British Isles and the Mediterranean world. It all began when Strachan was visiting Tel Gezer in Israel, an ancient site with a straight line of ten megalithic standing stones that are six to twelve feet tall. He remembered seeing a similar arrangement of stones from 3000 BC at Callanish in the Outer Hebrides. Both arrangements are oriented due north, a very difficult accomplishment, since there was no polestar or compass in those days.

Later he asked a guide what the village to the north was, and the answer was "Avalon."[54] This caused Strachan to research many connections between ancient Israel and Britain through the Pythagorean tradition, a brilliant mystery school that existed during the time of the Druids and Pythagoras. Eventually he was led back thousands of years into the megalithic time when Tel Gezer and Callanish were built.

Born around 580 BC, Pythagoras became very famous during the midpoint of the National Underworld in 550 BC. He was a master of mathematics and the numerical ratios that underlie the musical scale, and he traveled widely in Egypt, Mesopotamia, and Persia, where he mastered the ancient esoteric traditions from many teachers.[55] Under strict vows of secrecy administered by priests who were attempting to save precious preflood knowledge, Pythagoras was an oral transmitter of wisdom from earlier ancient civilizations. Whoever these ancient esoteric teachers were, they were right to adhere to vows of secrecy, since eventually the Romans and Christians torched much of their treasure stored in the Alexandrian Library in Egypt. Other than a few surviving fragments, before the Book of Enoch and the Qumran scrolls were found, Pythagoras and Plato were almost our only sources for preflood data.

To the Pythagoreans, numbers are first principles—the essence of material reality—that have existed since the beginning of time.[56] Platonic and Pythagorean philosophy have many similarities, and until the age of science, these ideas were compatible with Christianity. These ancient sacred traditions are being revived by the geometry and number ratios of the crop circles that are inspiring a modern awakening of sacred geometry. For example, by adding number-as-space to number-as-quantity, geometry emerges, which is a *theology of proportion* that links together the knowledge of all things. Eventually I discovered that Pythagorean thought is the basis of my nine-dimensional model, and that geometry exists eternally in 6D and guides proportion in 3D.

The Pythagorean teaching about number—called Gematria—is that letters are also numbers. The Gnostics—esoteric Christians who were condemned as heretics by the early Church—adopted Gematria as a central belief. With Gematria, if you know the number codes of the letters, you can read secret knowledge in coded writing, such as in the Bible, which is fun. The origins of this numerical alphabet go way back in time, and it shows evidence of highly intelligent planning, suggesting it is a preflood system.

Strachan uses Gematria to analyze Jesus's three names—Jesus, Christ, and Jesus Christ—and what he found is exceedingly thought provoking. Analyzing the names of Jesus (in Greek) with Gematria, three numbers come forth—888, 1,480, and 2,368. The *proportional relationship* between these numbers is 3 to 5 and 5 to 8, which is the Fibonacci scale or the Golden Section, the mathematical basis of creation in nature that can easily be seen in flowers and shells.[57] Strachan notes that, whether he realized it or not, "Jesus Christ embodied in his own names that principle of mean proportional, or mediation, which lay at the heart of his teaching about the reciprocal relationships between himself, God, and his disciples."[58] According to the nine-dimensional model, this would mean Jesus was in perfect alignment with his source, his 6D resonance. Then he pulled all the higher dimensions right into his body while being fully human—the Incarnation. As a rabbi, he would have been married and had a family while being a co-creator with God. The real point is that Christ came to reveal what each one of us can attain as Keepers of 3D: *The end of the Mayan Calendar is about each person moving back into proportional relationship with nature as Keepers of Earth.*

Of course, as we have seen, the early Church suppressed all this information. Also, for those readers who may feel that my thoughts

about Christ betray a Christian bias, the Church suppressed mor
the Gematria of the names of Christ. The Sufis—a mystical bra
Islam—report that Jesus did not die in the crucifixion. There are reliable stories of him ministering to Jews in Persia, Afghanistan, India, and Central Asia, a few years following the crucifixion. As well, Jesus is revered in the Quran, the Islamic scripture, which also says Jesus did not die in the crucifixion.[59]

I mention this to emphasize that Christ is much more universally beloved than most Christians seem to realize. I discuss Christ in this book since Quetzalcoatl is often thought of as Christ in the Mayan tradition. To consider that Christ represents the pinnacle of human evolution—as Strachan does by investigating the tradition that believed he was the Pantocrator—liberates him from the cross and from the Christian domination of his story. Strachan notes that many New Testament scholars assert that passages in the Bible that speak of "Jesus as the *Pantocrator*"—the creator of all things—are merely later assertions by "pious hagiographers" determined to divinize Jesus.[60] Yet the analysis of his Greek names by Gematria indicates that Christ *is* co-creator with God—Pantocrator. Using Gematria to analyze his names, he is identified with divine proportion—the symbolism of the Trinity and the Fibonacci number series. This places Christ in the realms of sacred geometry and sound, the essence of the Creator: *Christ as the living personified culmination of nature.* If it is true that divine light from Venus was used to bring in such a soul (possibly Enoch returned), then the same Venus alignment in AD 2001 suggests that the ancient rituals need to come back during our own times.

I am having fun with ancient wisdom, while we have one more new-paradigm writer to consider, Ian Lawton, who helped me finally answer why the Enochians are the Keepers of 9D. He also helped me imagine how the actual process of soul involution might work, and to investigate this process as the driver of evolution itself. Since Keepers hold the dimensions in form, it would make sense that some very high spiritual forces would be managing 9D—the highest dimension accessible to humans, which is also the Mayan Calendar.

ANTEDILUVIAN WISDOM IN *GENESIS UNVEILED*

Genesis Unveiled, by British esotericist Ian Lawton, is a whole book about the preflood human race.[61] Lawton believes (as I do) that this race was very spiritual but degenerated just before the cataclysm and/or during the survival period that followed. He thinks it is usually a waste of time to look for evidence of ancient advanced technology, since most of the remains of the preflood cultures were destroyed in the cataclysm and the rising seas.[62] Instead, he seeks to ascertain the spiritual nature of preflood people from the stories about them in ancient myths and literature. Using this approach, he has attained remarkable new insights about this originating spiritual race that adds more information to Calleman's evolutionary theories.

As we've seen, *From the Ashes of Angels* traces the Watchers back to around 8900 BC at Shanidar Cave, but Lawton goes back still farther in time and digs into some of the most intensely suppressed information that I know of. After all, if soul involution guided by divine consciousness is the real story behind human evolution, then there is no need for organized religion or the global elite and their armies.

In Genesis chapters 4 through 6, we are told Seth's descendants were originally a godly and simple race, but the descendants of Cain were materialistic and decadent. Because Cain's descendants prevailed over Seth's, God destroyed humankind in the Flood to cleanse the planet of this debased race.[63] Of course, this implies that the debasement of humanity caused the cataclysm, which is not a necessary conclusion at all. Earth changes happen, and the issue is whether evolution keeps on going. Until recently, Christian theologians maintained that the Flood was around 4000 BC; recent researchers, however, place it around 9500 BC. Additionally, there were other major floods after the global event, such as the 5600 BC Black Sea flood and a big localized flood in Sumeria around 4000 BC.[64] Unfortunately, as mentioned previously, these events are all jumbled up in the myths and epics that describe the great flood that almost destroyed the human race, and then humanity went through stages of degeneration from 9500 BC until about 6000 BC.

THE BOOK OF ENOCH

Early Judaism treasured the Book of Enoch because it contains preflood information that describes the human race before its degeneration. In reading the Book of Enoch, I've wondered whether the preflood race knew about our evolutionary past. Surely this would explain why the oldest sacred calendars are always based on 360 days. Calleman has shown how the Mayan Calendar is a record of the evolutionary timeline, the Vedic people protected their own preflood data in the Vedas, and there are parts of the evolutionary timeline in the preflood fragments. Did early Judaism venerate the Book of Enoch because they knew it was a record of its own evolutionary past? Imagine how precious this information would have been to them.

Just what did the preflood race know? If the preflood people actually knew about the great crawl from single-celled animals to complex spiritual humans, then is that why Enoch was said to have walked with God, that is, co-created with spirit?[65] Was Enoch, by his own willingness to incarnate, helping God know when humans were ready to evolve? Would we change if everybody knew about these remarkable evolutionary processes in time? What would life on Earth be like if we all realized that we walk with God every day of our lives as our cells, bones, and minds evolve?

The preflood race in the Book of Enoch were the Watchers, which means "awake," so who were they? We find some of their story in the seventh and eighth chapters of the Book of Enoch, which I will summarize: People on Earth (sons of men) had multiplied and had beautiful daughters. The Watchers (sons of heaven) beheld them, became enamored, and decided to select from them and bear children. Two hundred Watchers descended to Earth (even though their leader, Semyaza, was against it), and they cohabited with the women, who bore giants, the *Nephilim*. Once they grew up, the Nephilim turned against the people and devoured them, as well as animals, reptiles, and fish. After contact with the Nephilim, the people became sexually licentious; the Nephilim taught them to make weapons, and they also taught them the esoteric arts.[66]

This description of what the people became also has implications for what they were like *before* they degenerated. The fact that they began to eat animals implies that Earth's people were, at one time, vegetarians;

and since they had no weapons, at one point they had not been warlike. They must have forgotten the magical arts, so they were reinstructed in them. When the women gave birth, they brought forth giants, which is evidence for the difficult birthing described in Genesis as a curse of God. Knowing that the Book of Enoch is a combination of very old knowledge and later additions by translators who struggled to understand it (or change it), still there is a lot of information about what people might have been like before the 9500 BC cataclysm.

The term *Watchers* suggests that these angels had special awareness, and the Book of Enoch records that some of the Watchers and their leaders were opposed to breeding with the daughters of men.[67] Lawton proposes that these passages make sense only by means of a spiritual worldview—human advancement occurring through stages of soul incarnation.[68] Lawton says, "As part of the genus *Homo,* we must somehow have acted as a receptacle for progressively more advanced souls as we evolved. So what if at some point in our evolution, the human species appeared ready to receive at least some incarnating souls with a hitherto unmatched degree of karmic advancement?"[69] In other words, the two hundred Watchers may have been more advanced than the existing humans, which is also why some of them would not have wanted to breed with the daughters of men. After all, once the progeny—the Nephilim—were born, the children were anathema to both the angels and the humans.

Whatever the result was, Lawton proposes that such incarnations of advanced angelic souls would have been "the most explosive impact on human cultural evolution that it has ever received."[70] Regarding the puzzlement that anthropologists express about the hominid evolutionary advance 2 million years ago and the emergence of *Homo sapiens* one hundred thousand years ago (chapter 2), could there really be managed soul-infusion processes that trigger evolutionary advances during each new Underworld time acceleration? Are we in the middle of significant soul infusions during the Galactic Underworld? And what was it like for early human women to give birth to giants?

THE BIRTH OF GIANTS AND INCREASED CRANIAL CAPACITY

Consider this possibility: The birth of giants and the birthing difficulties described in Genesis and the Book of Enoch could have been caused by the problems that early hominids and humans experienced due to increasing cranial capacity. For example, in the evolutionary line that evolved into *Homo sapiens,* the average cranial size of *Homo habilus* (2 million years ago) is 600 to 750 cubic centimeters (cc); the cranial size of *Homo erectus* (1.5 million years ago) is 850 to 1,100 cc; and the cranial size of modern humans is around 1,350 cc.[71]

There are additional suggestions of difficulties on the evolutionary path in other preflood sources. When Mesopotamian scholars assembled various Sumerian and Akkadian texts during the nineteenth century (which they realized were preflood creation sources that predated Genesis), the story of the mysterious Anunnaki—who brought civilization to humankind—emerged. The Akkadian Atrahasis describes the creation of humanity, when the gods mix their intelligence with clay to make humans. Lawton wonders, is this a "veiled description of humankind receiving a relatively advanced soul for the first time?"[72] In the Akkadian *Epic of Gilgamesh,* there is a brief description of creating humans involving clay, and Enkidu is born who is the "offspring of silence," which means he can't speak. Also, Enkidu has shaggy hair, eats vegetation, and drinks water like a gazelle, which sounds just like a hunter-gatherer, or an early hominid.[73]

The Maya Popol Vuh describes the creation of different beings before the first human was created, and these passages have always puzzled me because they describe odd multiple attempts. Some have suggested these stages allude to what humans experienced during the Aztec world ages, but there is something else going on. The Popol Vuh says the Maker or Modeler fashioned various animals, but they were unable to speak to him or praise him. The Modeler tried many times to create humans working with mud, which is the universal mix in creation legends. The Maker went through various stages, but couldn't get a body to hold together or speak. Finally, he made manikins, or woodcarvings, but they didn't remember the Heart of the Sky. Then finally the Maker gave the creations heart, and they populated Earth, but still they did not speak. Then they were destroyed in a great flood, which is known to be the end of one of the world ages.[74]

The Popol Vuh informs us that monkeys look like people because they are *signs of a previous creation.*[75] This has to be a fragment referring to the 41 million-year-long Tribal Underworld, when monkeys slowly evolved into hominids! Preflood records may have contained the evolutionary databank, since some legends still have such odd fragments, but most of this precious memory was lost and suppressed. Lawton notes there is an emphasis on the early creations not being able to speak and pray to the gods, which would be a prerequisite for a golden race, or a race that could praise the Creator.[76] He says, "These more complex creation traditions just *may* be describing the idea that relatively advanced souls may have tried to incarnate in human form *before* our race was sufficiently advanced along the evolutionary path to make the experiment viable."[77]

This is a unique and thought-provoking view of the involution of spiritual beings into humans as we evolved. Many years ago, I asked the Pleiadians to take me back to the moment humans connected with God: I was a silent being who finally broke through and spoke to God, which felt like being a little child who speaks to his or her mother or father for the first time.

From my perspective, the soul involution of advanced beings would answer why the Enochians are the Keepers of the 9D. That is, the Pleiadians were saying that *the Enochians are the spiritual guides of the whole evolutionary process!* and the driver is the World Tree. Thus, when hominids were ready to become early humans, the Enochians began incarnating their own souls, just so that there would be beings on Earth that could recognize the Creator and praise the divine. Lawton calls this a "kick-start" in human evolution and goes back to one hundred thousand years ago for that "pivotal point in human development," which is, of course, the opening of the Regional Underworld.[78] Lawton notes that one hundred thousand years ago is when we have the first evidence for human burial as a ritual, which implies a feeling for the afterlife.[79] This connection to the afterlife is what is causing humans to evolve into co-creators with God.

The Popol Vuh has still more information about these creation processes. According to this ancient text, after several failed attempts, the goddess Xmucane ground yellow corn *nine times* and modeled our first mother-father. These first beings could talk and listen, they were seers and omniscient, and they had great knowledge, which is evidence of profound spiritual awareness.[80] This is a description of the creation of the *golden race,* the preflood race before it degenerated, which hints that *the golden race was nine-dimensional* (because Xmucane ground corn nine times).

However, for some reason, the gods felt the golden race was too good and was a potential threat, and so the gods took away their sight and knowledge, which must have been the debasement of a highly spiritual and omniscient creation.[81] Well, I think this description of creating humans and then limiting their powers is clearly a garbled attempt to explain the degeneration of humanity, which may be the most painful process that humankind has ever experienced. The loss of spiritual access must have happened due to the trauma of the cataclysm, and when people tried to tell this story, they had already fallen quite far, and they couldn't remember things. I actually believe that those of us who will be here on Earth in 2011 will again be omniscient seers, which is a return to our birthright.

In the Hopi creation stories, there are also stages when people cannot speak, that are more evidence of "difficulties encountered when advanced souls first attempted to incarnate in human or even protohuman forms."[82] Corroborating these difficulties, Berossus, a Babylonian historian-priest of the third century BC, who saved many extremely ancient fragments of early creation stories, described weird beings that were composites of humans and animals in Babylon, which are also described in many other traditions.[83]

Lawton summarizes these records of the creation of humanity: "I have postulated that this age [the golden age] arose as a result of the first angelic souls being able to incarnate in human form on earth in order to teach their human-souled fellows about the true nature of the ethereal as well as physical universe."[84] Once humans learned about the ethereal universe, they buried their dead to help them live in the ethereal universe. These beings in the ethereal—the Enochians—are the Keepers of all the soul infusion processes on Earth.

I end this book by adding my own thoughts about the golden age.

THE GOLDEN AGE AND THE BREAKTHROUGH CELEBRATION

The golden age was a time when humans were truly spiritual and in communion with the beings who created them. They had not yet experienced the cataclysm when Earth nearly died, and then during the long period of survival and pain, humans gradually lost the memory of the golden age. As their consciousness degenerated, they even thought they had angered God and caused the disaster.

The Enochians who loved them incarnated periodically to live as sages to assist struggling humankind to keep alive the memory of each person's inherent sight. Eventually, 5,125 years ago, human cultures began to really speed up. When the time was ripe at the midpoint of the National Underworld in 550 BC, very advanced souls such as Pythagoras, Zoroaster, Buddha, Solon, Lao-Tse, Isaiah II, Mahavira, Confucius, and many other sages we still remember today flooded Earth. Seeing that humanity was awakening, Isaiah announced that the divine proportion was coming—the Christ—as the Messiah, the anointed one. Once he was born, he joined with his female half and fathered a lineage. Through sixty generations his bloodline has infused humanity. Seeking the golden bloodline, people of all races have been intermarrying and giving birth to the Indigo Children, the Children of the Light.

Now during the Galactic Underworld, time acceleration is so intense that all the traumas and emotional blocks of the last 11,500 years are being transmuted through the actions of all humans during only a few years. The vortex for this transmutation is the Middle East, which is why the world is in such great pain now. Amid exploding bombs, millions remember the cataclysm that happened long ago, and soon the bombs will be swept away by the collective revulsion of all the people on Earth. Then in the last few years of the Galactic Underworld, people will deconstruct the religions that have blocked their access to the ethereal beings. As the religions fall, killing for God will cease, and war will end.

Beginning with the Breakthrough on Day Five of the Galactic Underworld—November 24, 2006—all humans will begin to respond to the call by the Universe society for guidance. Each person will want to know what each one of us must do as governments and religions crumble. During the Universal Underworld in AD 2011, the blood of Christ will quicken in the veins of all humanity, and love will be the greatest force on Earth. Mothers and fathers will refuse to send their progeny to war because their children are Christ. Wars will cease because people will see they don't need religion to be in communion with the divine.

The Keepers of Earth—humankind—will begin the long and hard task of restoring and reinvigorating Earth's ecosystems. We will see again that each plant and animal is infused with God's light. During the Universal Underworld in 2011—the Cosmic Party—all humans will exist in oneness with Earth's habitat, the womb of divine consciousness in the universe. During 2012, all the seasonal festivals, the equinoxes and the solstices, will be celebrated.

And when time ends and the evolutionary activation driven by the World Tree is finally complete, the people of Earth will have forgotten all about history and the Mayan Calendar; they will be in ecstatic communion with nature and the Creator.

APPENDIX A

REFLECTIONS ON EARTH'S TILTING AXIS

J. B. Delair's thoughts about why Earth's axis must have tilted 11,500 years ago from his "Planet in Crisis" article are next.[1] I also describe Alexander Marshack's research on Paleolithic and Neolithic bone markings as well as some very avant-garde Neolithic astronomical research by a few other scholars. This material could have been a whole chapter, but I chose to put it in an appendix because of its highly technical quality and because axial-tilt theory is a working hypothesis that I have by no means proven.

We begin with J. B. Delair:*

> The most immediately striking image of the Earth is that it rotates on an axis inclined at 23½ degrees from the vertical. Its orbit is not a perfect circle and it is not strictly concentric with the Sun. The axial tilt accounts for the variation in daylight hours per day between different parts of the world through the year. Combined with the Earth's eccentric year-long revolution around the Sun, the tilt accounts for the seasons and the difference between the average summer temperature North and South of the equator. Earth's axis of rotation does not coincide with its magnetic axis.

*The citations for the Delair article appear at the end of this appendix. See page 222.

This is also apparently connected with Earth's variable rotation, which fluctuates over a 10-year period.[28,29]

While following its orbit, Earth also oscillates cyclically: the "Chandler Wobble," with a cycle of 14 months.[30–32] This wobble is also associated with viscosity at the Earth's Core,[33] so it is an intimate part of Earth's present internal mechanism.

Because the Earth ought, theoretically, to rotate on a vertical axis and may actually have done so in the geologically recent past,[34,35] *these details suggest a planet which, in not very remote times, has been seriously disturbed* [italics mine]. If true, these "ill-fitting terrestrial cogs" [i.e., Earth's inner core rotates significantly faster than the rest of the planet], which appear to function only through the presence and action of inner-Earth viscosity, may be regarded as abnormalities.

Yet a number of these features, including axial inclinations and eccentric orbital paths, are shared by several of Earth's planetary neighbors, so are the terrestrial equivalents really "normal" ones? The evidence suggests otherwise. A selection of this evidence and its pan-solar ramifications, has been discussed by Allan and Delair.[36]

There ought to be no reason why any Earth-like planet undisturbed for untold ages should not have a vertically positioned axis [italics mine]. This would unify the locations of the geographical and magnetic poles, ensure equal daylight hours in all latitudes and virtually eliminate the seasons. There would be no necessity for various subcrustal layers to function differentially and the present rheological mechanism would not be required. However, an equatorial bulge would remain as an essential stabilising feature and the retention of a non-circular and non-concentric orbit would probably still cause small "seasonal" climatic differences when Earth was nearer or further from the Sun.

What follows next in "Planet in Crisis" is a few comments about the implications of these abnormalities, and then Delair describes the Holocene Earth changes. He notes that these relatively recent global Earth changes must be the result of other forces operating deep within Earth, and he speculates on the "cause or causes of the Holocene cluster of catastrophes" in this section, which is titled "Rupture."

Earth's essential instability, as mirrored by its structural and behavioral "abnormalities" must reflect some persistent internal imbalance not yet fully accounted for. However we have already seen that the boundary of the solid Mantle with the liquid outer Core is irregular, perhaps to the point of being topographic,[119] and that the outer surface of the solid inner Core is also not smooth.[120] Indeed, it is uncertain whether the inner Core is actually spherical: as it moves within the viscous medium of the outer Core it need not necessarily be. A non-spherical or irregularly surfaced inner Core would, however, generate further instabilities.

Given these details and the inner Core's higher rotational speed [four or five hundred years for one complete turn by the inner Core!], its surface irregularities must be in continuously varying opposition to those of the slower moving Mantle's inner boundary. The plastic material of the intervening outer Core must therefore undergo displacement as the distances between the opposing irregularities alter. Continuous compression and release of this material must occur as the outcome of such differential rotation. Inner Earth movements of this kind are now being investigated deductively.[121]

It is not unreasonable to infer that peaks of acute outer Core compression and troughs of compensatory relaxation should alternately develop to differing subsurface intensities at different times in different hemispheres. Likely results, which could sometimes arise quite suddenly, would include events which have often been termed mid-Holocene catastrophes. Lithospheric adjustments such as the Scandinavian, Alpine and South American lake tilts, extensive regional subsidences such as the Indonesian/Australasian/Melanesian area, large-scale water table changes as in the Arabian and Saharan regions and earthquakes and severe vulcanism, such as the Santorini eruption and its widespread aftermath effects,[122] are typical examples. There is also no doubt that large earthquakes such as at one time racked the Roman Empire are closely associated with the Chandler Wobble,[123,124] itself intimately connected with viscosity activity at the Earth's core.[125]

Why does the inner Core rotate faster than the rest of the planet and why doesn't the inclination of its axis coincide with that of Earth as a whole (cf. the different locations of the geographic and magnetic poles)? *It strains credulity to suppose that any Earth-like*

> *planet, undisturbed by external influences for millions of years, could have naturally acquired, unaided, a tilted axis, an offset magnetic field, variable rotation, or a Chandler Wobble* [italics mine].
>
> Most geoscientists who have studied this broadly agree that any event or series of events resulting in characteristics as profound as these would almost certainly have to involve some influential outside agency. In other words Earth would need to be subjected to some powerful extraterrestrial force—a force severe enough to rupture its previous internal mechanism without actually destroying it.
>
> Down the centuries precisely such a source has been repeatedly advocated to account for traditionally catastrophic events like Noah's Deluge, the loss of a primaeval Golden Age, the advent and also the demise of the Ice Age, the sudden refrigeration of the Siberian/Alaskan mammoth fauna and even the foundering of legendary realms such as Atlantis, Lyonesse, etc.[126–135]

Next the article discusses the main catastrophic scenario and advances these events as the cause of Earth's abnormalities, such as the axial tilt.

> After apparently adversely affecting many of the Sun's outer planets, the postulated cosmic visitor was seemingly able to temporarily retard the rotation of Earth's Mantle and lithosphere but could not halt the rotation of the inner Core, due to the viscosity of the outer Core. As a consequence of this disruption, Earth's thermal and electromagnetic levels increased enormously, with all kinds of unwelcome effects. Among these appear to have been an *axial slewing of the Mantle and crust to an inclination differing from that of the solid inner Core* [italics mine]. Indeed, the latter may itself have been wrenched, gravitationally, within the liquid outer Core to an off-centre position, causing the Earth to yaw or tremble (or both), as some traditions recalling the events actually state. Such movements were *only* possible because of the viscosity of the outer Core. It is also probable that the cosmic assailant pulled the entire Earth over to its present inclination, since any former *normal* planetary regime *must* have developed over a more vertical axis.
>
> The resumed rotation of the Mantle/Lithosphere around a still rotating but slight off centre inner Core (the liquid outer Core is

> immaterial at this point) at different speeds round different axes (the geographical and magnetic poles) imposed huge strains and stresses on the Earth. Prominent were fluctuating rotation and the Chandler Wobble. An off-centre Core would ensure only a very slow and stuttering return to planetary normality, punctuated sporadically by catastrophic terrestrial adjustments. Holocene history is littered with these. While often alarming, they are really the coughs and wheezes of a world still in crisis.

Catastrophobia is the psychological syndrome resulting from these Earth changes over 11,500 years. By remembering the original event and recognizing the valiant adjustments to the new Earth by the Neolithic people, we can heal this syndrome. The memories of these catastrophes were saved because the ancient people knew that their ancestors—all of us—would need this information to be able to achieve the next stage of our evolution. This is Turtle Medicine. Recently, some amazing new theories about ancient astronomy have been advanced that highlight how Neolithic humans came to terms with the new Earth. I will cover some of these new theories very briefly, since these new ideas may end up being fertile ground for others who may want to consider whether Earth's axis may have tilted recently and made life on Earth into a whole new ball game.

Science writer Alexander Marshack was asked in 1962 by leaders in the space program to co-write a book that would explain how humankind had reached the point of planning a Moon landing. When Marshack interviewed many of the key movers and shakers in the space program, he realized none of them knew *why* we were going into space; all that mattered was they had the skills to do it. He was supposed to write a few pages on the dawn of civilization and how the development of mathematics, astronomy, and science leads up to entering space. He studied the orthodox interpretation of our historical emergence and ran right into all the "suddenlys" that begin ten thousand years ago, such as the instant-flowering model of Egyptian civilization.[2]

He went back into Paleolithic cultures, and then he had the big awakening that changed his whole life, as well as our current understanding of Paleolithic and Neolithic science: Marshack discovered he could read and decode the markings by early humans on ancient bones. Ironically, he was working on a project that was to explain how we could get to the

Moon, and he discovered that Paleolithic and Neolithic carved bones are lunar calendars! Eventually he became intrigued by the fact that from the earliest times until 9000 BC (Early Neolithic), the bones are lunar calendars, and then about ten thousand years ago, the solar factor is added to the lunar notations. Suddenly, the lunar phases are divided into six-month phases, which suggests that these phases begin with either an equinox or a solstice.[3] *Marshack's research and the research of many other paleoscientists indicates that early humans show no signs of being aware of the existence of the four seasons until ten thousand years ago.* I think it is unrealistic to surmise they just didn't notice sunrises and sunsets traveling back and forth on the horizon and the changing seasons, especially since they suddenly became obsessed with this factor approximately ten thousand years ago.

Marshack's laborious decodings of the bone markings as lunar calendars from before 10,000 BC have been widely accepted over the past forty years by most prehistorians.[4] Eventually, he focused on the fact that the bone notations were lunar calendars until the end of the Paleolithic and then added the solar factor ten thousand years ago. He spent twenty years trying to decipher the marking of a plaque found in 1969 at the Grotte du Tai that is ten to eleven thousand years old, because it has the typical lunar cycles with some new elements. Marshack decoded it by using all his knowledge of Upper Palaeolithic notations and art combined with the art and notations of Neolithic preliterate cultures. This plaque is one of the earliest and most complex scientific Early Neolithic objects, and it is probably one of the first objects that records attempts by humans to show that there are about six lunar cycles between the solstices and equinoxes.[5] Anthropologist Richard Rudgley notes that the Tai notation fills the vacuum "before the apparently sudden development of astronomical observations in the Neolithic period in north-western Europe, epitomized by alignment of megalithic monuments, such as Stonehenge."[6] Regarding Early Neolithic astronomical notations, what strikes me the most is the wall art of Çatal Hüyük, the complex geometrical Natufian designs, and the incised spirals and chevrons of Newgrange, which show the year divided into the light and dark halves with the phases of the Moon delineated.[7]

We have only barely begun to realize how advanced Neolithic astronomy was because only now are we deciphering their monuments, notations, and artifacts. It is very hard for us to take their obsession with the sky seriously because we can barely see the night sky in our modern cities. A great change is evident in Neolithic renditions of the sky, and I believe it was the tilting of the axis that caused this change. I think we can suddenly see things that were right under our noses because our own perspective is expanding. For example, Robert Temple, the author of *The Sirius Mystery,* published a brilliant book, *The Crystal Sun,* that definitively proves people have been using telescopes and lenses for improving eyesight for thousands of years, and many of the lenses have been on view in museums all over the world for hundreds of years.[8]

Ralph Ellis, author of *Thoth,* makes a strong case that Avebury Circle is a representation of Earth floating in space. Also according to Ellis, *Avebury Circle exhibits Earth's axial tilt!*[9] The diagrams and text by Ellis need to be studied by interested readers. As for me, in the 1980s, I noticed that the north/south avenues coming into Avebury tilt about 23 degrees to the east/west avenues, and I wondered why. Why would they go to so much trouble to model Earth in space tilted in its solar orbit? Well, megalithic astronomy exhibits a virtual obsession with the solstices and equinoxes. For example, Newgrange captures the first light of the winter solstice when the Sun sends daggers of light deep into its chambers to illuminate the centers of complex spirals. Many other megalithic chambers capture the light at the exact moment of the spring or fall equinox. Even the center of the Vatican is constructed to capture the spring equinox light. (I noticed this orientation when I visited the Vatican in 1979.)

In *Uriel's Machine,* Christopher Knight and Robert Lomas have shown how early cultures went to almost unbelievable lengths to comprehend, record, and anticipate light from the Sun and Venus. They have demonstrated that the shapes of various lozenges depicted on Neolithic Grooved Ware pottery and incised stone balls actually convey astronomical information. The shape of the lozenges created by sunrise and sunset through the year changes with latitude; they believe these lozenges depict the latitude of the makers![10] Suddenly latitude became important because the solar angles by season change so dramatically in the northern latitudes.

The wild theories in *Uriel's Machine* need much testing, and the authors' original conclusions about the megalithic stone chamber Bryn Celli Ddu on the Isle of Anglesey (3500 BC) merit serious attention.

The authors demonstrate by archaeoastronomy that Bryn Celli Ddu is a sophisticated chamber that was used to correct the time drift in solar and lunar calendars by calibrating the time of year by the eight-year Venus synodic return cycle with the winter solstice. Venus shines a bright dagger of light every eighth year into the chamber at Bryn Celli Ddu when it is the brightest. According to the Roman historian Tacitus, this was when the Goddess appeared, and since Venus is the most accurate indicator of the time of the year, what does time have to do with the Goddess?[11]

In *The Dawn of Astronomy,* Sir J. Norman Lockyer reported in his exhaustive study of the star temples of ancient Egypt that various temples are aligned to certain key stars as far back as 6400 BC. He also demonstrated that the "apertures in the pylons and separating walls of Egyptian temples exactly represent the diaphragms in the modern telescope," and comments that they "knew nothing about telescopes."[12] Robert Temple has subsequently demonstrated that the ancient Egyptians *did* have telescopes, so perhaps their temples were used like the large telescopes in modern observatories.[13]

I bring together these related details here because I believe that *the tilting axis inspired a preliterate scientific revolution* that we are decoding in our times. The axial tilt changed the way we receive light on Earth. Alexander Marshack was asked to find the source of humanity's ability to get to the Moon, and he discovered that archaic people were already profoundly in touch with the Moon in their times. I suggest the way beyond catastrophobia is to awaken this archaic intelligence, which I find is encoded in the light changes caused by the solar angles (moving sun on the horizon) that shift according to latitudinal degrees—the Light. According to indigenous traditions, the Light is infused with cosmic information, and modern science has discovered that photons carry cosmic information. Megalithic astronomy, as well as indigenous astronomy, suggests that the Light is more potent and transmutative for humans during the equinoxes, solstices, and new and full moons. Perhaps that intentional attunement awakens cosmic intelligence. Perhaps a new evolutionary form began when the tilting axis cracked Earth open, as if Earth were a cosmic egg ready to hatch in the universe.

NOTES FROM J. B. DELAIR'S ARTICLE:

28. Lambeck, K, *The Earth's Variable Rotation: Geophysical Causes and Consequences* (Cambridge, 1980).
29. Rochester, MG, *Phil.Trans.Roy.Soc.Lond.,* vol. A306, 1984, pp. 95–105.
30. Ray, RD, Eames, RJ & Chao, BF, *Nature,* vol. 391, 1996, n. 65831, pp. 595–597.
31. Dahlen, FA, *Geophys.Journ.Roy.Astron.Soc.,* vol. 52, 1979.
32. Guinot, B, *Astron.Astrophys.,* vol. 19, 1972, pp. 207–214.
33. Ibid.
34. Harris, J, *Celestial Spheres and Doctrine of the Earth's Perpendicular Axis,* Montreal, 1976.
35. Warren, RF, *Paradise Found; The Cradle of the Human Race at the North Pole. A Study of the Prehistoric World,* Boston, 1885, p. 181.
36. Allan, D. S., and Delair, J. B., *When the Earth Nearly Died,* Bath, 1995. [This is the British edition of *Cataclysm!,* Santa Fe, 1997].
119. Keaney, P., ed., *The Encyclopedia of the Solid Earth Sciences,* Oxford, 1993, p. 134.
120. Whaler, K & Holme, R, *Nature,* vol. 382, no. 6588, 1996, pp. 205–206.
121. Ramalli, G, *Rheology of the Earth,* 2nd edn., London, 1995.
122. Pellegrino, O, *Return to Sodom and Gomorrah,* New York, 1995.
123. Mansinha, L & Smylie, *DL, Journ.Geophys.Res.,* vol. 72, 1967, pp. 4731–4743.
124. Dahlen, FA, *Geophys.Journ.Roy.Astron.Soc.,* vol. 32, 1973, pp. 203–217.
125. Yatskiv, YS & Sasao, T, *Nature,* vol. 255, no. 5510, 1975, p. 655.
126. Whiston, W, *A New Theory of the Earth,* London, 1696.
127. Catcott, A, *A Treatise on the Deluge,* London, 2nd ed., 1761.
128. Donnelly, I, *Ragnarok: The Age of Fire and Gravel,* 13th ed., New York, 1895.
129. Beaumont, C, *The Mysterious Comet,* London, 1932.
130. Bellamy, HS, *Moons, Myths, and Men,* London, 1936.
131. Velikovsky, I, *Worlds in Collision,* London, 1950.
132. Patten, DW, *The Biblical Flood and the Ice Epoch,* Seattle, 1966.
133. Muck, O, *The Secret of Atlantis,* London, 1978.
134. Englehardt, WV, *Sber.Heidel.Akad.Wiss.Math.Nat.KL.,* 2 abh, 1979.
135. Clube, V, and Napier, WR, *The Cosmic Serpent,* London, 1982.

APPENDIX B

ASTROLOGICAL TRANSITS THROUGH AD 2012

ASTROLOGICAL SUPPORT FOR TIME-ACCELERATION THEORY

The Galactic Underworld (January 5, 1999, through October 28, 2011) is a cycle that processes the development of civilization during the National Underworld (3315 BC through AD October 28, 2011) and the rise of technology during the Planetary Underworld (1755 through October 28, 2011). During the Galactic, by means of time acceleration everything goes so fast (twenty times faster than during the Planetary) that beliefs and needs of the past transform themselves into new ways of being. Since there will be a final twenty-fold acceleration during the Universal Underworld (February 11, 2011 through October 28, 2011), there is no doubt this process will culminate. Of course, imagining things going twenty times faster during 2011 than they have been going since 1999 is almost impossible.

During the National, power and control systems were developed through theocracies and various political systems that are all now changing. During the Planetary, industry and technology connected all the people on the planet, yet these systems also are changing. Now, with the advent of the Galactic in 1999, the evolutionary agenda is *human enlightenment*—a time when the people come into harmony with nature; all activities and beliefs that separate the people from nature must go out of form. This is a time when all political control systems and technological systems that exploit life in the universe must go.

We humans live on a planet that is one among many others orbiting the Sun. Our Sun is a star that swims around the center of the Milky Way Galaxy like a dolphin of light, diving up and down through the Galactic plane. As an astrologer, I have seen that the locations and aspects of the planets express huge archetypal patterns that influence human behavior and the Sun. Astrology teaches that the qualities of the planetary fields (as well as other factors in the universe, such as the Sun, Moon, and Galaxy) influence human behavior. For example, Venus encourages human attraction, Mars inspires us to express our power, and Jupiter shows us how to find comfort and abundance. Knowing about these archetypal forces makes it possible to live life with more awareness and foresight, or to identify in hindsight why things worked out as they did so we learn from our mistakes.

My personal astrological work since 1991 has been *macroastrology,* which analyzes very large patterns. My Web site—*www.handclow2012.com*—explains how the planetary forces influence political systems as well as our lives within these systems. As revealed in this book, based on Calleman's research, I'm convinced the Mayan Calendar *does* track Nine Underworlds of evolution in the universe that have been accelerating by factors of twenty beginning 16.4 billion years ago. I am also convinced, being an astrologer, that if humankind is to attain enlightenment during the Galactic Underworld, the astrological patterns should show how the great archetypal forces might facilitate this process. Astrology is based on the Hermetic principle—"As above, so below"—so, if astrology *is* a real influence on human feelings and behavior, the planetary patterns would actually describe the attainment of enlightenment from 1999 through 2011. So let's take a look.

Previously (in appendix A of *The Pleiadian Agenda*) I described the astrological transits from 1972 through 2012. Upon reviewing that appendix, I found my earlier analysis to be accurate and useful; however, it covers a longer period than the Galactic Underworld. We now have the advantage of knowing what actually did happen between 1999 through the summer of 2006, which is more than half of the Galactic Underworld. The earlier forecast was good guidance at the time, but now we have many more facts to go on. Also, in those days nobody was discussing time acceleration by factors of twenty, or that incremental jumps would occur on January 5, 1999, and again on February 11, 2011. I can now look back at 1999 through 2006 to see how the astrological patterns functioned during this period of remarkable acceleration that began in early 1999.

Why do I say *"remarkable acceleration"?* In this book, we have already seen that the galactic alignment of 1998 coincided with detectable physical changes in Earth and the universe. Now that the galactic acceleration is rolling along, we are interested in seeing how astrological patterns are enabling humankind to achieve the apotheosis—enlightenment—in a few short years.

THE OUTER PLANETS AND HUMAN ENLIGHTENMENT

The current shift in human consciousness toward enlightenment actually began with Day Seven of the National Underworld (1617 through 2011), which was, and is, the *fruition phase* of the National. Since Day Seven opened in 1617, three outer planets have been discovered—Uranus in 1781, Neptune in 1846, and Pluto in 1930—as well as critically important Chiron (which orbits between Saturn and Uranus) in 1977. Uranus's discovery coincided with the discovery of electricity, and Neptune's with heightened spirituality in everyday life, as expressed through the rise of Spiritualism and Transcendentalism during this period. Pluto is the most unyielding transformative force in the solar system, and its discovery happened as fascism and communism took hold in Europe and Russia, and the nuclear bomb was invented. Chiron, the wounded healer, helps us see how personal wounding triggers us to evolve and forces us to face the most painful hidden places in our souls. We need the wounded-healer process because it is impossible to attain enlightenment without surrendering to inner darkness and making space for cosmic light.

In general, the outer planets—Chiron, Uranus, Neptune, and Pluto—rule transpersonal states of consciousness; they prod us to remember that *enlightenment is the reason for living*. The innermost astrological influences—Mercury, Venus, Moon, Mars, Jupiter, and Saturn—create constant change just like daily weather patterns. Inner bodies will not be analyzed unless they are involved in major configurations with the outer planets. Jupiter, because it is large enough to be a star, often has great influence in our personal lives, so I always watch for its participation in outer planet configurations. Saturn structures the outer-planet archetypes, which inspires them to make real patterns in our lives. Saturn is thus looked at very carefully to assess how we can attain human enlightenment. We must all learn to utilize Saturn's discipline; it teaches each one of us how to attain very high states of mind. We cannot access the high-energy aspects of Chiron, Uranus, Neptune, and Pluto transits without Saturn's influence. If planetary forces are real, then if Saturn were not playing a major role, attaining enlightenment beginning in 1999 would be hard to imagine. I love Saturn, and I love how Chiron moving between Saturn and Uranus (also sometimes inside Jupiter) breaks down our emotional blocks, so we can utilize the wild, transformative, electrical Uranus to access the powers of Neptune and Pluto.

THE WILD AND CRAZY 1960s

To set the stage for the drama—1999 through 2011—we have to go back in time to gather some background. We need to comprehend the existing energetic field that was in place in early 1999. Just as a theater's stage has art and props that create a specific atmosphere for the drama that is to be played out once the curtain rises, transits during the wild 1960s set the stage for the Galactic Underworld. Uranus came together with Pluto (conjunct within less than 5 degrees) during 1964 through 1968, which initiated a time-release depthcharge of radical transformation in culture and individuals. Uranus rules change and transformation, and Pluto rules processing our deepest and darkest emotions. In modern times, Uranus and Pluto are always in key angular aspects when radical shifts of consciousness occur. When they come together, the greatest forces of change are set in motion. Since Pluto was only sighted in 1930, the 1960s saw the first-known conjunction of Uranus and Pluto. Before this, we had not been conscious of the human potential for enlightenment of all people, only of an occasional guru or saint.

Going back to the 1960s in your mind, do you remember the Summer of Love in San Francisco, the Beatles, and the Beatnik generation? Since then, the media and the previous generation have done everything possible to make it look like all that crazy behavior just went away, but it has not. Knowing how expanded consciousness feels has caused this cultural seed to germinate within the planetary mind ever since the 1960s. In fact, the youth of the 1960s manifested all the aspects of enlightenment—meditation, art, personal searching, love of nature, a yearning for peace, and cosmic fusion by means of altered states of awareness.

According to astrology, whatever happened during the conjunction of Uranus and Pluto in the 1960s would manifest globally as soon as these two planets reached their *first square* (a 90-degree angle) to each other. For example, each month I analyze the qualities of the New Moon on my Web site, and I watch to see if the special energetic patterns I described for the New Moon become visible at the lunar first quarter, the Moon's releasing square to the Sun. Usually they do, and if they do not, I go back to the drawing board to see what I missed. During each lunar cycle, everything germinates during the first seven days and then becomes visible during the first square.

With Uranus and Pluto, when the first square comes, the creative explosion of the 1960s will release into the world. Guess what? *Uranus comes to its close square to Pluto during 2011,* and then the exact squares are during 2012 on June 24 and September 19! Just coincidence? Well, the exact reading for the Uranus/Pluto squares in 2011 through 2013 is that *the enlightenment energy discovered by the Children of Love will manifest as a global force in 2011, and during 2012, no one who lives on the Earth will be able to resist melting into enlightenment.* You will live and breathe enlightenment. I was one of the Children of Love living in San Francisco during the late 1960s and I experienced the waves of enlightenment. I can assure you that you will not be able to resist being swept into joy, creativity, and fusion with awesome cosmic forces; it would be like resisting a tsunami.

Another major influence during the 1960s was Saturn transiting Aquarius from 1962 to 1964, when the first vibrations of the coming Age of Aquarius were felt. This emerging energy was exemplified in the musical *Hair,* which set the stage for the Uranus/Pluto creative explosion. Saturn was in Pisces from 1964 to 1967 during the Uranus/Pluto conjunctions, and this Piscean structural phase supported a spiritual awakening that again emerged from 1991 to 1996, when again Saturn transited Aquarius and Pisces. I mention these two Saturn cycles of the 1960s because Saturn in Aquarius and Pisces (1991 to 1996) was a practice run for Uranus in Aquarius (1996 to 2003) and in Pisces (2003 to 2011); and for Neptune in Aquarius (1998 to 2012).

During the 1962–1967 and 1991–1996 "Saturn practice runs," profound cultural and spiritual changes were going on, such as the study of Eastern mysticism in the West and the new-paradigm questing movement described in this book. As we've seen, these movements are formulating a program for enlightenment to be directed by Uranus in Aquarius and Pisces (1996 to 2011) and Neptune in Aquarius (1998 to 2012). Why? As you already know, Uranus rules radical transformation of self and cultures; well, *Neptune rules enlightenment processes.* Neptune dissolves our resistance to the spiritual forces in the universe and erodes egoic boundaries until we surrender to bliss. Actually, the most important planet to track from 1999 to 2012 is Neptune, since it rules all the processes of enlightenment. Neptune in Aquarius from 1998 to 2012 brings in the coming Aquarian age of human enlightenment.

The only way we can actually master Neptunian spiritual enlightenment is to transform our emotional limitations by embracing absolute truth and integrity, which is Pluto's agenda. Pluto is the key to finding this emotional purity, since it rules transformation of inner darkness so that cosmic fusion and galactic integration are experienced. Meanwhile, Pluto has a very elliptical orbit, and it goes inside Neptune's orbit for about twenty years during each of its 249-year orbits around the Sun. Pluto was inside Neptune's orbit from 1979 until the spring equinox in 1999. While inside Neptune's orbit, Pluto's radical transformative power was intensified for us, while Neptune was out there in the outermost reaches of the solar system communing with the universe. Just a few short months after the Galactic Underworld opened, Pluto moved out beyond Neptune, and Neptune began to dissolve our boundaries to spirituality again. That is, we went through radical psychological transformation from 1979 to 1999, and then spirit began again penetrating our psyches honed by Neptune. Maybe we needed this short break from Pluto to get ready for transforming the cultural patterns of the National and Planetary during only 12.8 short years. Yet, I wonder, how *can* Pluto accomplish the transformation of 5,125 years of endemic cultural patterns?

CHIRON, PLUTO, AND THE GALACTIC CENTER

When the Galactic Underworld opened January 5, 1999, Pluto and Chiron were both in Sagittarius. When Day One of the Galactic was complete on December 31, 1999, Chiron was conjunct Pluto—sending a potent message on the last day of the twentieth century: *Chiron conjunct*

Pluto in 11 Sagittarius is the signature aspect of the New Millennium as well as the completion of Day One of the Galactic. Day One is the day of sowing the Galactic Underworld creation; thus, the planetary archetypal force that planted the enlightenment seeds is Chiron and Pluto in Sagittarius, the zodiacal location of the Galactic Center.

It is interesting to me that Chiron and Pluto have shared a similar fate in the hands of today's revisionist astronomers. When Chiron was sighted in 1977, astronomers declared it a planet, so astrologers went to work to determine its influence. We noticed that Chiron's influence is extraordinary for its size: just like tiny homeopathic remedies, this small body has a huge influence on people. Then, sometime during the 1990s, astronomers decided Chiron was merely a planetoid, or maybe a comet or asteroid. Many astrologers, including myself, still consider it to be a major influence, and most astrologers will continue to use Pluto as a potent element in people's charts as well. It's really weird that present-day astronomers are trying to take Pluto's archetypal force away from the people by downgrading it to a dwarf planet, especially since it even has a moon, Charon. Pluto was the god of the deep thousands of years ago, so surely he has not just gone away. Are the astronomers uncomfortable with the exploration of human emotional depth that has been going on since the 1930s?

Chiron rules transmutation of inner emotional blocks by encouraging people to access the deepest levels of wounding—the experiences that created the blocks in the first place. On a more subtle level, reading Chiron correctly reveals when an individual first split out of divine awareness and got caught in the merely physical sense of self, which happens to almost everybody. We are much more than just physical beings, yet people cease feeling cosmic connection and Earth becomes a prison. When Pluto has done his work forcing us to face the dark side and change deep levels of resistance, then Chiron kicks in to help us see where we are disconnected from the divine. Regarding this painful truth about ourselves, if the cause was not our own personal pain (which is rarely the case), then Chiron forces us to clearly see the truth about man's inhumanity to man, which could instantaneously change the world.

For example, if the majority of Americans really saw the pain that weapons manufactured in the United States cause in the world, there would be a massive uprising against supporting this economy. If we can just recognize the moment we lost divine connection, then we can build a bridge back to our greater sense of self. Chiron conjuncted the Galactic Center during 2001, signaling that a galactic understanding about human wounding would begin. These aspects between Chiron and Pluto during Day One of the Galactic signal that *we will process the limiting patterns of the Planetary and National Underworlds by facing our own inner pain*. What could that possibly be? What might it mean?

TRANSMUTING THE BATTLE BETWEEN EAST AND WEST

As I complete this book, we are in the second half of Night Four of the Galactic, and it is easy to see how this process is working. The world is in the middle of a fast-acting and tragic World War III–level battle between the East and the West that first developed during the National Underworld, and then became murderous once technology was used to improve weaponry during the Planetary Underworld. Horrible aggression, violent defensiveness, and intransigent behavior on the part of insensitive world leaders and religious fanatics is exposing the insanity of 5,125 years of wars in the name of God; the technology of death forces each one of us to look deep within seeking to clear our own inner darkness. The desecration of the human body in our times is unbelievable, thus I call our time the Age of Flying Body Parts. Yet, as people see the awesome scope of evil in the world, it becomes increasingly untenable to live by the programs of the National and Planetary underworlds; only an entirely new human can transmute these old ways. I have no doubt those who survive will change, because Chiron in late Aquarius conjuncts Neptune in Aquarius in 2009 to 2010, which means we will move our personal spiritual wounding to the greater whole and surrender violence. We are at a loss without contact with true spirituality; we've been cut off from the greater aspects of *being* for a long time.

By means of astrology, we can look at various stages of planetary aspects from 1999 through 2012 to imagine what might forge the new human. This involves looking at the angular relationships between Saturn, Chiron, Uranus, Neptune, and Pluto during 1999 through 2012, as well as adding other planets when they aspect these bodies. For those of you who are wondering why I include 2012, since the time acceleration is complete in 2011, astrology shows that 2012 is about learning to utilize the acceleration of the Universal. You may recall from the text that an alignment (closest in 1998) of the rising winter solstice Sun with the Galactic Center is going on during the end of the Calendar. This greater access to galactic energy has been aligning our consciousness with the Galaxy itself, therefore I include aspects to the Galatic Center in this analysis. The Galactic Underworld is the critical period of this adjustment to cosmic frequencies, and then the winter solstices of 2011 and 2012 will enable humans to integrate the vastly intensified galactic frequencies.

Possibly the most astonishing thing about the dance of the outer planets during the end of the Calendar is that during 2012 Uranus squares Pluto, which means the countercultural enlightenment released during the 1960s will once again flood the planet. The point is, *time acceleration during the Universal Underworld in 2011 will be when the form of the new spiritual human coalesces, and 2012 will be when we first learn how to live as new spiritual humans.*

In 1998, Neptune went into Aquarius, the sign it was in when it was first sighted in 1846—a time when idealistic spiritual movements came forth in America. Thus, Neptune completes its first orbit during the Galactic Underworld. Once we have experienced the full orbit of a new planet, the archetype of that planet is then fully integrated into the human psyche. As Neptune completes its first-known solar orbit in 2011, the energy that spawned the nineteenth-century utopian movements will come forth again, especially when Neptune goes fully into its home sign, Pisces, in 2012.

Uranus was first sighted in 1781 during the beginning stages of the Planetary Underworld, which has greatly influenced the development of technology during this Underworld. Pluto was first sighted in 1930; thus we are just barely getting used to the intense psychological influence of this god of the deep. Pluto was in its home sign, Scorpio, from 1983 to 1995, which means we have experienced the most intense levels of Pluto just before the end of the Calendar. Pluto in Scorpio (also inside Neptune's orbit) prepared the way for intense spiritual work by intensifying the emergence of the deep subconscious; it forced profound emotional exploration for many people. Pluto passed into Sagittarius in 1995, which pressured us to seek truth and personal integrity.

Great integrity has been maturing beautifully into spiritual genius during Pluto in Sagittarius from 1995 to the present, and will continue to do so to 2008. It has been the main influence in exposing the outrageous dishonesty and lies of the Roman Catholic Church, which hijacked the truth about the life of Christ for 1,500 years, as already discussed in the text. Pluto conjuncts the Galactic Center in 27 Sagittarius during 2006 and 2007, meaning the cosmic Christ will infuse the world with light. As it leaves the Galactic Center, Pluto then moves into squares to the equinox and solstice degrees—0 degrees Aries, Cancer, Libra, and Capricorn—as it moves into Capricorn and inspires a new spirituality. Pluto squaring the seasonal quarters during 2008 through 2010 will greatly intensify the personal manifestation powers of the seasons, which means the vast majority will experience their own relationship with God and will abhor organized religion. During the winter solstices, Pluto will continue to intensify human connection with the Galactic Center right during the time of year when the birth of Christ is celebrated.

As you can easily see, everywhere you turn in the solar system through 2012, extraordinary astrological synchronicities will be going on with the outer planets and the Galactic Center. These configurations and placements are exactly what we need to handle the rapid time acceleration of the Galactic and Universal. Of course, if we humans are to attain something as profound as enlightenment, we would need more prosaic structuring and disciplinary forces to get us to exceed our human limitations. These forces do exist during the end of the Calendar, and they are also the main reason life has been so difficult for all of us since 1999. This is why the philosophers have always noted it is difficult to live during significant times.

THE GREAT SQUARES IN THE SKY DURING THE GALACTIC UNDERWORLD

The way to find these structural disciplinary forces is to look for "stressful" aspects between the outer planets that also sometimes involve some inner planets that "tie" us to transpersonal forces. Stressful aspects are conjunctions (0-degree angles); squares (90-degree angles); and oppositions (180-degree angles). These are the angles that try our souls and also bring out our greatness. One extremely stressful configuration is the *T-square* when one or more planets form a square to two or more planets that are in opposition. Another is the *grand square,* when four or more planets are in squares (90-degree angles) that involve two oppositions.

Conjunctions, squares, and oppositions can occur in *cardinal signs*

(Aries, Cancer, Libra, and Capricorn) that start processes; in *mutable signs* (Gemini, Virgo, Sagittarius, and Pisces) that widen and explore the processes; or in *fixed signs* (Taurus, Leo, Scorpio, and Aquarius) that complete and fix the processes. *Fixed grand squares* cause monumental changes in the world. Like the four horsemen of the Apocalypse, they force individuals to release separateness and enter into the flow of life, in this case, into cosmic consciousness. Fixed grand squares involving the outer planets are exceedingly rare (occurring only in thousands of years). Take note: *fixed grand squares involving the outer planets are the big feature of the Galactic Underworld.* Therefore, I must decode these patterns, even though they can be complicated for most readers. These patterns are so extraordinary and long-lasting that nothing like this has occurred since the Calendar was invented. These configurations do have the potential to culminate evolution so that the "new human" can be born.

Going back to August 1999, during Day One of the Galactic, a fixed grand square culminated during a solar eclipse in 18 degrees Leo—Mars in Scorpio was opposite Saturn and Jupiter in Taurus, squared by the New Moon in Leo opposite Uranus in Aquarius. This signified we were resolving personal conflicts between the will (Leo) and the higher mind (Uranus), while involved in a struggle between materialism (Taurus) and the search for emotional depth (Scorpio). The various aspects of this grand square moved in and out of placement during the summer of 1999, and the tension was palpable and very hard to handle. I was writing and teaching a lot about this grand square while watching people close at hand, and I was amazed by the stubbornness and ego that people brazenly displayed. I saw relationships and friendships shatter while personal confidence went down the drain. And in U.S. politics, President Bill Clinton was being vilified to pave the way for the Bushites. Calleman says the Galactic Underworld is the Apocalypse, which has turned out to be prophetic. When I taught students about this amazing grand square, I didn't realize it was the beginning of the Apocalypse, but it was.

Meanwhile, during 1999 we were also all feeling that time was accelerating twenty times faster; things felt out of control. Then, on the very last day of Day One, Chiron conjuncted Pluto, and we began processing the deep pain issues of the National and Planetary underworlds by facing hard truths. Looking at what the sky was saying in the summer of 1999, it makes sense that a cabal of destroyers—the

Bushites—would decide to run a global empire and tear apart existing systems in less than thirteen years! From that point of view, these ecological and emotional trashers may be the natural facilitators of such monumental change.

Day Two of the Galactic—December 25, 2000, to December 20, 2001—was the germination phase of the Galactic, and its signature event was the bombing of the World Trade Center, which is the key symbol of globalism. The astrology of 9/11 is so predictive that it is scary. Saturn, which rules structures and forms, opposed Pluto, which rules deep emotions and radical transformations. The forces that hold reality together—and those that tear it apart—were in radical conflict and outright hostility. The deep anger against the West that was building in the Middle East became visible, and the attack showed that America was not invincible. The fall of the Twin Towers also revived subconscious memories of wildness and chaos when Saturn opposed Pluto previously in 1965 and 1966. Saturn opposite Pluto causes deep unease and tension that forces previously hidden issues out in the open. These world-altering events during the germinating phase of Day Two signaled that East/West tension would be the central theme of the Galactic, yet polarization normally tends to move toward resolution. During the Saturn/Pluto oppositions of the mid-1960s, the idea of enlightenment came to the West by means of Eastern spirituality, mostly from India and the Orient. At that time, the controlling forces in the West were very suspicious of these ideas. During the oppositions of Pluto and Saturn in 2001 and 2002, we can see that the Islamic people believe they are the enlightened ones on the planet now; they will not back down in the face of what they consider to be great evil.

During Day Three of the Galactic (December 15, 2002, to December 10, 2003), the sprouting phase of the Galactic occurred. This was the year Uranus went into Pisces (where it will be until March 2011), which made the spiritual aspects of change and transformation more visible, exemplified by the appearance of Dan Brown's sales-record-shattering novel, *The Da Vinci Code,* in 2003. Uranus in Aquarius, and then in Pisces, during the end of the Calendar means that spiritual transformation is the point of everything—and will keep growing until the end.

During Day Four of the Galactic (December 4, 2004, to November 29, 2005), the real spread or proliferation of the movement into enlightenment was happening. The world stage was polarized, with America sinking into a quagmire in Iraq; yet many more people were feeling profound spiritual forces that were more important than the events in the news. The great tsunami in Indonesia on December 26, 2004, right after the opening of Day Four, was the most powerful global uniting of human need and love in human history; *compassion had greater value than aggression and violence.* Chiron went into Aquarius in early 2005 and will eventually join up with Neptune in Aquarius during 2009 through 2011. Many people could feel themselves change inside as they contemplated the suffering in the world, or they were caught right in the middle of the action in the various theaters of war. During Day Four, many great teachers were appearing on the world stage, and students were realizing that they themselves are teachers.

During the end of Day Four in late 2005, we all experienced another fixed grand square, which annealed our souls into more complete spiritual vehicles. During November and December 2005 and January 2006 (when Night Four began), Mars in Taurus (materialism) opposed Jupiter in Scorpio (deep emotions) and was squared by Saturn in Leo (will and ego) opposite Neptune in Aquarius (higher ideals). This fixed-sign square moved in and out of aspect for around three months, since Mars began this pattern in retrograde motion and then went direct. This signaled a titanic struggle over personal needs and making a better world. During these squares, T-squares, and grand squares during late 2005 and early 2006, many people felt overcome with meaningless as they saw their cherished belief systems collapsing all around them. Americans lost faith in their government when it failed its citizens during and after the hurricanes Katrina and Rita, and when it wasted billions of dollars, and many thousands of lives, in wars. Since these belief systems come from the National and Planetary Underworld structures, a collapse had to come. However, few people could imagine what would replace the old ways of life, and they just became depressed. The grand square was remarkably enduring and tight, and few people escaped its tearing wrath in their lives. Night Four was a winter of deep discontent that was actually deepening people, if they could learn anything from it.

THE FALL OF THE SYSTEMS OF THE NATIONAL AND PLANETARY UNDERWORLDS

During 2006, Pluto nearly conjuncted the Galactic Center, which inspired deep levels of galactic integration, and Saturn moved into exact opposition to Neptune during late August. When Saturn opposes Neptune, old systems fall because it is time to align planetary rules and laws with higher spiritual potential. For example, the Berlin Wall fell in 1989 when Saturn conjuncted Neptune, which was a signal that this Saturn/Neptune cycle would be about the fall of fanatical and controlling systems. During the Saturn/Neptune opposition (the fruition of the 1989 conjunction) in late August 2006, what seems to be falling apart is Western influence in the Middle East. The United States is stuck in a quagmire in Iraq and Afghanistan, and Israel is being criticized for inflicting so much damage and loss of life in Lebanon and Gaza.

Whatever happens, great realignments of power and influence are developing in the Middle East, since that is where the three great religions—Judaism, Christianity, and Islam—were spawned during the National Underworld. Actually, I think *organized religion is what is falling,* because as people become more spiritual and less political, it becomes impossible to kill for God.

In other words, during the human alignment to greater spirituality inspired by Saturn opposite Neptune, the faults of organized religions are too much for people to tolerate. In the middle of all the conflict of 2006, another series of fixed-sign T-squares formed from August to September: Saturn in Leo (will) opposed Neptune in Aquarius (higher spiritual potential) and squared Jupiter in Scorpio (unprocessed dark neediness). The fall of 2006 devolved into a titanic battle over whose god was right. This struggle is ominous in light of great East–West tension, but it seems to be necessary to release old National and Planetary belief systems. The point is, the people have to withdraw their support for the battles, especially Western support for invading and occupying sovereign nations of the East. The people of the West need to realize the reason they are in danger is because of the aggressive policies of their own governments, and they need to see that these policies will never succeed because making countries in the East into enemies is very dangerous; after all, the West has invaded the East, not vice versa.

During Day Five of the Galactic (most of 2007), Pluto transits the

Galactic Center, which will inspire the development of more universal points of view. The great squares will have released, but Saturn opposes Neptune again during February and June of 2007, ensuring the continual collapse of old National and Planetary systems and the realignment to greater spirituality. As this happens, there will be an ongoing crisis of belief that builds and builds in the major religions. Many people will see that these religions have been the major cause of wars during the last 5,125 years, and they will take back their hidden (subconscious) emotional support of the killing. However, nasty dogs do not go away without biting, so this time will not be pleasant. Great tension and violence between East and West is highly likely. The world will be very changed once this opposition is complete by summer 2007, when new alignments based on higher principles are being formulated. During 2007, David will be much favored as he confronts Goliath, because peak oil will make it more and more difficult to run war machines.

THE FINAL PHASE OF THE GALACTIC UNDERWORLD: 2008–2011

Meanwhile, the planets are getting ready to cook up a new crock of soup during 2008–2011 for the final phase of the Galactic Underworld. This time Saturn, Uranus, and Pluto will be in various squaring and oppositional formations as they move into *cardinal signs or initiating processes*. Saturn will oppose Uranus periodically from November 2008 through July 2010, and Saturn will square Pluto periodically from October 2009 through September 2010. Then Uranus will square Pluto from 2011 through 2015, which carries its potent force right to the end of the Calendar and beyond. During August of 2010, Saturn in Libra will oppose Uranus in early Aries, while Pluto in Capricorn T-squares this opposition. Cardinal signs initiate changes, which will feel very different from the locking effect of the fixed squares. There will be a lot of fast, dizzying changes (many of them economic, based on energy problems) that will be very hard to cope with unless you are debt-free and really aligned with spiritual forces. When Saturn opposed Neptune in 2006 (as it will again in 2007), many new alignments to higher spiritual forces developed, and after 2008 these new spiritual choices need to be a real part of your life, not just dreams.

With Saturn opposite Uranus, huge transformations of existing structures will occur. Because Uranus is in Aries and Saturn is in Libra, old structures will be transformed in shocking ways that will create balance within new structures. For example, this opposition could force America to reduce its energy consumption and curb its tendency to take fast actions without thinking about the consequences. As Pluto in Capricorn squares Saturn in Libra, the structural forces (Saturn) will be transformed amid revolutions in the actual structure of the universe (Pluto). The big struggle during this balancing and integration will be the most aggressive level of change and transformation (Uranus in Aries) that can be imagined. Yet these are the kinds of archetypal forces that are needed to push people to choose enlightenment over materialism.

During 2011, Neptune completes its first solar orbit since it was sighted, and when Neptune enters Pisces, the forces of Spiritualism that developed around 1846 will reemerge. For example, American idealism—a movement that once inspired the world—will express itself through a new Transcendentalism that will be inspired by galactic consciousness. Politics based on the universe (exopolitics) may replace politics based on America-as-empire. Everywhere there will be contempt for militarism and global chauvinism. Uranus was conjunct Pluto during the 1960s, which introduced enlightenment, and with Uranus arriving at its first square to Pluto during 2011 and 2012, there is no doubt enlightenment will sweep our planet as each country offers its own spiritual contribution.

The astrological patterns in play during the Galactic Underworld demonstrate that the planets support the theorized time accelerations of the Galactic and Universal underworlds. The small planet the astronomers demoted in 2006 tells the story all by itself: during January 2008, Pluto goes into Capricorn for the first time since it was sighted. Capricorn is the maximum structural energy force on Earth, thus Pluto in Capricorn will confront and radically destroy the remaining elements of the National and Planetary underworlds. The most profound energies available on Earth—the ability to be enlightened and in joy—will emerge as the highest creation of the Galactic and Universal Underworlds in 2011.

APPENDIX C

GUIDE TO THE GALACTIC UNDERWORLD

Figure C.1—Guide to the Galactic Underworld—can be used many different ways. The first and most important thing is that you photocopy the unmarked figure and set aside a clean copy, because you may want to use it in different ways.

The figure is a thirteen-storied pyramid of the Galactic Underworld with the opening days of each Day and Night indicated, as well as the date for the midpoint of Day Four, the apex of the Galactic Underworld time acceleration.

So, assuming you have an unmarked copy to work with, take that copy to a photocopy machine, blow it up double or triple size, and make sure you have lots of blank space to the left and right of the pyramid for your own notes. This will work on an 8½-by-14-inch page, with the majority of the free space to the right and left of the pyramid. Once you blow it up and have allowed the blank space, you may want to make two more copies, so that you have three enlarged copies. Now you are ready to go!

Exercise A. Working with your enlarged copy, you may want to draw lines off the corners of the pyramid that delineate the opening date of each Day and Night. You want thirteen clearly delineated spaces for your personal notes. Next, try to remember what you were doing in 1999, 2000, 2001, and so on. You can indicate big historical events, such as 9/11 during 2001, because often this helps you remember what you were doing around that time. You may want to put historical events in red ink and personal events in blue. Write these notes concisely, because you will find you remember more and more as you work with this. It's a great time to drag out your calendars, diaries, and notes, and talk to your family and friends about what you all were doing.

Once you've gathered enough data, you may notice several things: (1) your life changed radically during 1999, and by 2000, you were already wondering what on Earth was going on; (2) things appeared in 2001 that were first created during 1999; (3) you were rather lost and bummed out during 2000, 2002, and 2004. You may see that themes that came through during 1999, 2001, and 2003 exploded during 2005.

What's the point of all this? First of all, doing this exercise will improve your understanding of the Galactic acceleration, which has been very confusing for most people. Second, as far as I can tell, what we each created and processed from Day One through Day Four represents *major transformation themes for the rest of the Calendar*. Based on my personal experience with the exercise, it might be wise for you to decide not to open up any more major thematic files until 2012, and to consciously complete all the themes that came up during 1999 through 2005. As

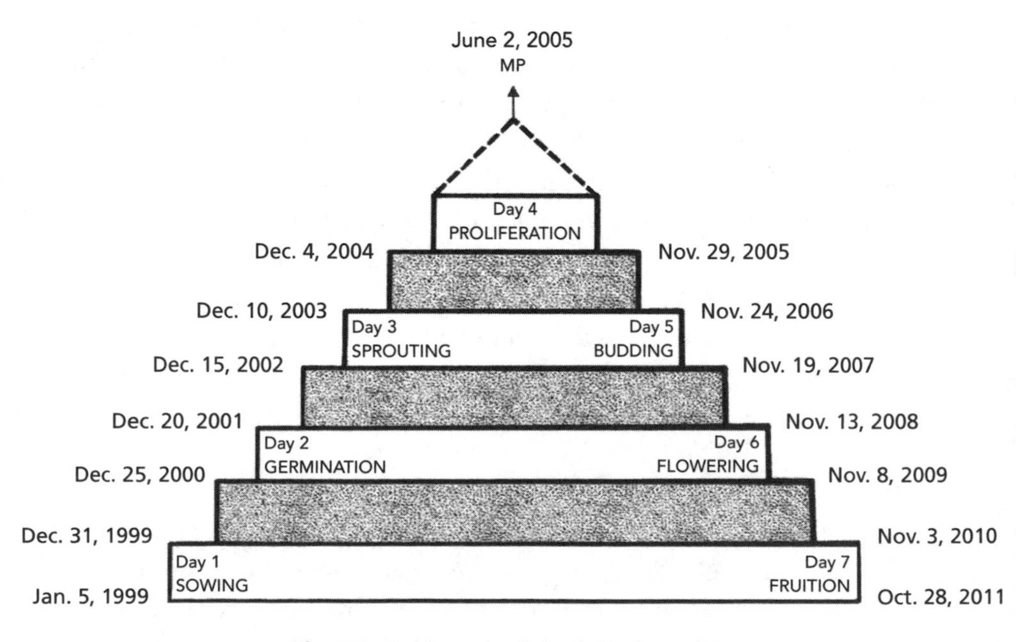

Fig. C.1. *Guide to the Galactic Underworld.*

you already know—if you have made some progress with remembering what you've been doing—what has been going on in your life is almost too much, yet it is also mostly very valuable. It is time to harvest your new garden during Day Five through Day Seven, not a time to plant new seeds. That way, whatever your unique gift is at this time, it will mature to be shared with everyone on the planet.

Exercise B. Take another enlarged copy and draw in your lines to delineate the Days and Nights. Then go to appendix B and study the astrological transits of the Galactic Underworld and perhaps take some notes on them. Once you have a grip on what they are, mark them into the correct Days and Nights. You may even want to diagram a key configuration, such as the grand squares of August 1999. If you are a budding astrologer, put in the transits described in appendix B, and then get out your ephemeris and mark down some more of your own. I was limited by space, but you are not! Look for lunar eclipses, for example, or write down how the transits aspect your own natal chart and try to remember what was going on during those transits. Have fun!

APPENDIX D

FINDING YOUR MAYA DAY SIGN

Conversion Codex for the Mayan Calendar

Calculate your day sign according to the true Sacred Calendar Count used by the Maya for 2500 years. This Conversion Codex was created by Ian Lungold and is reprinted here with permission.

INSTRUCTIONS

1. Find your year of birth in the Year Table below. Write down the number below your year of birth. (Note! If you were born in the months of January or February, use the year prior to your year of birth from the Year Table. Example: If you were born in January 1949, please calculate it as if you had been born in 1948.)
2. Find your month and day of birth in the Month and Day Table (see page 244). Write down the number in the Month and Day Table that corresponds to your date of birth.
3. Add the numbers from #1 and #2 above. If you were born before sunrise, you should subtract the number one from the sum of the two. If this sum is less than or equal to 260, move on to #4. If the sum is greater than 260, then subtract 260 from it and move on to #4.
4. Find the number between 1 and 260 that you arrived at in #3 in the lower part of the rows in the Sacred Calendar (see pages 245–247).

The number within the parentheses above this is your cosmic tone. In the column on the left just above that row, you will then find your Maya day sign.

YEAR TABLE

1910	1923	1936	1949	1962	1975	1988	2001
49	**117**	**186**	**254**	**62**	**130**	**199**	**7**
1911	1924	1937	1950	1963	1976	1989	2002
154	**223**	**31**	**99**	**167**	**236**	**44**	**112**
1912	1925	1938	1951	1964	1977	1990	2003
260	**68**	**136**	**204**	**13**	**81**	**149**	**217**
1913	1926	1939	1952	1965	1978	1991	2004
105	**173**	**241**	**50**	**118**	**186**	**254**	**63**
1914	1927	1940	1953	1966	1979	1992	2005
210	**18**	**87**	**155**	**223**	**31**	**100**	**168**
1915	1928	1941	1954	1967	1980	1993	2006
55	**124**	**192**	**260**	**68**	**137**	**205**	**13**
1916	1929	1942	1955	1968	1981	1994	2007
161	**229**	**37**	**105**	**174**	**242**	**50**	**118**
1917	1930	1943	1956	1969	1982	1995	2008
6	**74**	**142**	**211**	**19**	**87**	**155**	**224**
1918	1931	1944	1957	1970	1983	1996	2009
111	**179**	**248**	**56**	**124**	**192**	**1**	**69**
1919	1932	1945	1958	1971	1984	1997	2010
216	**25**	**93**	**161**	**229**	**38**	**106**	**174**
1920	1933	1946	1959	1972	1985	1998	2011
62	**130**	**198**	**6**	**75**	**143**	**211**	**19**
1921	1934	1947	1960	1973	1986	1999	
167	**235**	**43**	**112**	**180**	**248**	**56**	
1922	1935	1948	1961	1974	1987	2000	
12	**80**	**149**	**217**	**25**	**93**	**162**	

MONTH AND DAY TABLE

DAY	JAN	FEB	MAR	APR	MAY	JUNE	JULY	AUG	SEP	OCT	NOV	DEC
1	46	77	0	31	61	92	122	153	184	214	245	15
2	47	78	1	32	62	93	123	154	185	215	246	16
3	48	79	2	33	63	94	124	155	186	216	247	17
4	49	80	3	34	64	95	125	156	187	217	248	18
5	50	81	4	35	65	96	126	157	188	218	249	19
6	51	82	5	36	66	97	127	158	189	219	250	20
7	52	83	6	37	67	98	128	159	190	220	251	21
8	53	84	7	38	68	99	129	160	191	221	252	22
9	54	85	8	39	69	100	130	161	192	222	253	23
10	55	86	9	40	70	101	131	162	193	223	254	24
11	56	87	10	41	71	102	132	163	194	224	255	25
12	57	88	11	42	72	103	133	164	195	225	256	26
13	58	89	12	43	73	104	134	165	196	226	257	27
14	59	90	13	44	74	105	135	166	197	227	258	28
15	60	91	14	45	75	106	136	167	198	228	259	29
16	61	92	15	46	76	107	137	168	199	229	260	30
17	62	93	16	47	77	108	138	169	200	230	1	31
18	63	94	17	48	78	109	139	170	201	231	2	32
19	64	95	18	49	79	110	140	171	202	232	3	33
20	65	96	19	50	80	111	141	172	203	233	4	34
21	66	97	20	51	81	112	142	173	204	234	5	35
22	67	98	21	52	82	113	143	174	205	235	6	36
23	68	99	22	53	83	114	144	175	206	236	7	37
24	69	100	23	54	84	115	145	176	207	237	8	38
25	70	101	24	55	85	116	146	177	208	238	9	39
26	71	102	25	56	86	117	147	178	209	239	10	40
27	72	103	26	57	87	118	148	179	210	240	11	41
28	73	104	27	58	88	119	149	180	211	241	12	42
29	74	105	28	59	89	120	150	181	212	242	13	43
30	75		29	60	90	121	151	182	213	243	14	44
31	76		30		91		152	183		244		45

THE SACRED CALENDAR

Alligator

(1)	(8)	(2)	(9)	(3)	(10)	(4)	(11)	(5)	(12)	(6)	(13)	(7)
1	21	41	61	81	101	121	141	161	181	201	221	241

Wind

(2)	(9)	(3)	(10)	(4)	(11)	(5)	(12)	(6)	(13)	(7)	(1)	(8)
2	22	42	62	82	102	122	142	162	182	202	222	242

Hearth

(3)	(10)	(4)	(11)	(5)	(12)	(6)	(13)	(7)	(1)	(8)	(2)	(9)
3	23	43	63	83	103	123	143	163	183	203	223	243

Lizard

(4)	(11)	(5)	(12)	(6)	(13)	(7)	(1)	(8)	(2)	(9)	(3)	(10)
4	24	44	64	84	104	124	144	164	184	204	224	244

Serpent

(5)	(12)	(6)	(13)	(7)	(1)	(8)	(2)	(9)	(3)	(10)	(4)	(11)
5	25	45	65	85	105	125	145	165	185	205	225	245

Death

(6)	(13)	(7)	(1)	(8)	(2)	(9)	(3)	(10)	(4)	(11)	(5)	(12)
6	26	46	66	86	106	126	146	166	186	206	226	246

Deer

(7)	(1)	(8)	(2)	(9)	(3)	(10)	(4)	(11)	(5)	(12)	(6)	(13)
7	27	47	67	87	107	127	147	167	187	207	227	247

Rabbit

(8)	(2)	(9)	(3)	(10)	(4)	(11)	(5)	(12)	(6)	(13)	(7)	(1)
8	28	48	68	88	108	128	148	168	188	208	228	248

Water

(9)	(3)	(10)	(4)	(11)	(5)	(12)	(6)	(13)	(7)	(1)	(8)	(2)
9	29	49	69	89	109	129	149	169	189	209	229	249

Dog

(10)	(4)	(11)	(5)	(12)	(6)	(13)	(7)	(1)	(8)	(2)	(9)	(3)
10	30	50	70	90	110	130	150	170	190	210	230	250

Monkey

(11)	(5)	(12)	(6)	(13)	(7)	(1)	(8)	(2)	(9)	(3)	(10)	(4)
11	31	51	71	91	111	131	151	171	191	211	231	251

Road

(12)	(6)	(13)	(7)	(1)	(8)	(2)	(9)	(3)	(10)	(4)	(11)	(5)
12	32	52	72	92	112	132	152	172	192	212	232	252

Cane

(13)	(7)	(1)	(8)	(2)	(9)	(3)	(10)	(4)	(11)	(5)	(12)	(6)
13	33	53	73	93	113	133	153	173	193	213	233	253

Jaguar

(1)	(8)	(2)	(9)	(3)	(10)	(4)	(11)	(5)	(12)	(6)	(13)	(7)
14	34	54	74	94	114	134	154	174	194	214	234	254

Eagle

(2)	(9)	(3)	(10)	(4)	(11)	(5)	(12)	(6)	(13)	(7)	(1)	(8)
15	35	55	75	95	115	135	155	175	195	215	235	255

Owl

(3)	(10)	(4)	(11)	(5)	(12)	(6)	(13)	(7)	(1)	(8)	(2)	(9)
16	36	56	76	96	116	136	156	176	196	216	236	256

Earth

(4)	(11)	(5)	(12)	(6)	(13)	(7)	(1)	(8)	(2)	(9)	(3)	(10)
17	37	57	77	97	117	137	157	177	197	217	237	257

Flint

(5)	(12)	(6)	(13)	(7)	(1)	(8)	(2)	(9)	(3)	(10)	(4)	(11)
18	38	58	78	98	118	138	158	178	198	218	238	258

Rainstorm

(6)	(13)	(7)	(1)	(8)	(2)	(9)	(3)	(10)	(4)	(11)	(5)	(12)
19	39	59	79	99	119	139	159	179	199	219	239	259

Light

(7)	(1)	(8)	(2)	(9)	(3)	(10)	(4)	(11)	(5)	(12)	(6)	(13)
20	40	60	80	100	120	140	160	180	200	220	240	260

THE DAY SIGNS

Alligator (Imix)

Alligator is the initial day sign and often manifests in people as someone who initiates new projects. It is an Eastern energizing day sign that uses very powerful instincts to bring up new phenomena and creative ideas from the deep flow of the collective unconscious. Usually completing the projects that surface, however, is not their forte. For this reason it is essential that they cooperate with others in order to have productive results. Alligators usually also have a caring side with a strong nurturing and protective energy. They care about their offspring and may work hard to provide security for family and friends, and they should take care not to become overprotective and dominating.

Wind (Ik)

Wind refers to spirit and breath, and this is one of the clearly spiritual, and sometimes even airy, day signs. It is often in need of grounding. Wind is a Northern sign and so has a certain quality of cool detachment. Wind embodies the ability to communicate and disseminate good thoughts and ideas. Wind people are dreamers with strong imaginations and make excellent teachers and journalists. They may be very good speakers and spread spiritual inspiration "with the wind." Like the wind itself, these people are extremely flexible. This flexibility may cause them to be very indecisive and they may seem, to others, to be too fickle and incoherent. Wind people may be destructive, both to themselves and others, when they assume a lofty air. If they find a natural way of being grounded, Wind people will be very inspiring to others.

Hearth (Akbal)

This day sign can be given either of the names: Hearth, House, or Night. It is said that there is a special kind of feminine softness in the people that embody this sign. The meaning of House and Hearth also has to do with protecting one's own home and family and keeping away the powers of the night. Imagine a person sitting by the hearth at the center of the house who is telling legends and fairy tales. The carriers of this sign may, among the Maya, become shamans through their acquaintance with the darker regions of the human psyche. They may help dispel uncertainties and doubts that were born out of the darkness or the subconscious mind. Hearth persons can find new solutions and artistic inspiration through their knowledge of the emptiness of the Night. Yet, unless they courageously travel into the dark, they may very easily find themselves in doubt and insecurity.

Lizard (Kan)

Sensualism or sensuality is characteristic of this day sign, which is also sometimes translated as Net or Seed. All of these connotations contribute to its meaning. It is a Southern sign, and this sun-exposed character contributes to both its sensual and somewhat leisurely style. The Lizard was considered to be the sign that controlled the sexual force of the body. As natural networkers, Seeds strive to liberate themselves and others from the oppressive patterns of the past, and it is with this intention that the Seed person may plant the new seeds—or is one with him/herself. To create true prosperity, the Lizard must learn to appreciate all the gifts that he/she receives and conduct profound inner investigations.

Serpent (Chicchan)

As in so many other traditions, the Serpent represents magical powers. This means that it may be a very spiritual sign, but with powers that easily may be abused. It is an intelligent Eastern sign that may be electrifying with its new incoming energy. Through their sincere willingness to serve others, Serpents may be open to expand their hearts. This sign holds authority. Serpents are very flexible, or even fluid, until they are caught in a corner where they may explode. Serpents with a poisonous temper may poison themselves and others by creating oppressive or even destructive attitudes.

Death (Cimi)

There is almost invariably a special kind of softness about people born on this day sign. Traditionally, Death was in the ancient sources considered as the luckiest of days, and supposedly Death people may be very successful in business. They may, however, also be excellent healers who calmly guide others through the transformations of life because of their own spiritual strength. Death people support pregnant people and guide them through the transition to motherhood; they also have an ability to pass themselves through lasting transformations at key periods in their lives. Among the Maya, life was generated by death and the contact with the ancestors served to activate inherent psychic abilities. The challenge for Death people is to live life fully and to not give in to defeatism.

Deer (Manik)

Deer is one of the truly spiritual day signs, but it would be a mistake to identify it with a timid, gentle deer. Rather, it should be likened to the powerful stag. Even if they embody a peaceful day sign, Deer people may indeed express their power very directly. Yet, they do so with a strong concern for the welfare of others and respect of the spirituality present in all. Deer are dominators that protect and sacrifice their own interest for others. This desire for dominance—which is not always apparent immediately to others—sometimes makes their relationships complicated, causing them to taken on unconventional forms. When confronted with promises it has not kept, Deer becomes stubborn, manipulative, and evading. Their challenge in life is to be able to balance the power of the Deer with humbleness and understanding for others.

Rabbit (Lamat)

The energy of the Rabbit is growth and attraction of "pure luck." It is a Southern sign and therefore is characterized by a certain ease in its life. If Rabbit is just willing to surrender to life, things will come easily. This day sign is associated with the fertility of the rabbits through their ability to makes things multiply, and in general this is a day sign that will enjoy growth whether it is about money or having "a green thumb." Rabbit's natural tendency to be drawn toward harmony and ease may, however, also become compulsory and make Rabbit too pleasant and too generous. When a Rabbit thinks it has given too much, it gets weakened and may even collapse. Rabbits are not tough people, and the challenge for them is to create a strong core in their center from which their power may emanate.

Water (Muluc)

Water is a fairly difficult day sign to be born into. It has an intensifying Eastern energy that makes it hard for the strong emotions that are evoked by it to be contained. For those with this day sign, emotions may flow over in all directions without control. Water people are imaginative and have psychic abilities and an inclination for performance. They are certainly not rigid in their thinking. This means not only are they spontaneous, they sometimes feel others cannot understand them. Others will sense hidden motives in Water and may find it hard to trust them. Water people may evoke strong sexual and violent feelings in those that they come into contact with. They are, however, also often very charming and fun to be with, and this is what is behind their willingness to perform.

Dog (Oc)

Dogs are loyal, enduring, and goodhearted. People born into this day sign are seen as warm, alert, and brave, and if they find the right mission in life to which they can be loyal, they may accomplish very great things. Dogs like to be parts of teams and are often leaders. However, they usually do not create the causes they work for themselves. Those that are born under this day sign may be very sensual and know how to enjoy life. Among the Maya, the Dog is considered a strongly sexual sign and paradoxically, despite their spiritual loyalty, they are considered prone to sexual escapades. Dogs are ambitious and will take the opportunities as they manifest, including in the form of infidelity. Among the Cherokee, the day sign of Dog was Wolf. The challenge of Dog is to find a task in life where its many good qualities—such as goodheartedness and endurance—may be used most beneficially.

Monkey (Chuen)

Among the Maya, the Monkey is known as the weaver of time. Because it is the day sign in the middle, Monkey is often very creative, knowing many different things and being able to bind these together. Monkeys may be much fun to be with; they are charming, sometimes uncontrolled and bent for practical jokes. They are jacks-of-all-trades and typically have artistic abilities. There is also an annoying side to this charm, since they often have a compulsory need to be in the center of everyone's attention. Some may even act stupid to get into the limelight. Usually they have a short attention span and are ready to move on to something else and may never stay to learn to master anything fully. They, however, know how to do a lot of things and are never boring to be around.

Road (Eb)

Sometimes this day sign is also called Grass. Road persons are much devoted to their fellow humans. They care about the community at large, about future generations, and the children of the Earth. Yet, they have little or no desire for recognition or standing in the limelight for the good that they do. They may often keep a low profile and are soft and considerate in their ways. The Southern character of this day sign creates a certain ease about their lives. Many will take care of the poor, sick, and elderly and make personal sacrifices for these groups. This compassionate side they have also means they are easily hurt. They are easy to like, devoted, and hardworking, and therefore have success in business and travels. If they encapsulate negative feelings and disappointments that are not expressed, however, they may easily be afflicted by disease or a poisoned view of others.

Cane (Ben)

Those that are born into the day sign Cane—sometimes also called Reed or Staff—have a special relationship to Quetzalcoatl, the god of light and duality, and are thus considered as very worthy. They are usually recognized by their style of authority that is symbolized the by Staff of Life. Among the Maya, this is a symbol of the spiritual authority of an elder. In fact, the name Reed for this day sign is fairly misleading as they are usually not fragile at all. Instead, they are often the carriers of a fair amount of authority and they know it. Cane is an Eastern sign and so its carriers have a strong collected energy. Those born into this day sign are often leaders in society and as parents they are considered to be wise. They will often fight for a cause that they have found worthy. However, Cane people need a lot of appreciation and because of their inflexible views and high expectations, they do not easily become intimate with others. Thus, they may also have problems in marriage and business. Their challenge in life is to develop their flexibility so that the Reed is not broken.

Jaguar (Ix)

The Jaguar is a secretive being that sneaks around in the night. This is the typical day sign for prophets among the Maya and because of their ability to see in the dark, those with this sign often have clairvoyant ability paired with a fair amount of intelligence. This is a very typical day sign among day-keepers. The intellectual powers of the Jaguar combined with its typical feminine streak may create a healing ability in its carriers. Their strength has given them patience, which can rapidly shift into quick runs of activity. There is, however, often a very narrow focus of their endeavors and they are rarely open to explore alternative paths of life. They tend to be difficult for others to get a hold of, and they tend to just turn up and disappear into other people's lives. Their relationships are often complicated and dominated by the theme of eating or being eaten, and their secretiveness does not make it easier to live with them. Their challenge in life is to humble and open to other people.

Eagle (Men)

Eagle people are both powerful and ambitious, and they have high aspirations for their lives. There is a soaring energy about them and life may occur to them as a constant dream flight, as symbolized by their totem animal. Often they are successful in achieving material abundance and fortune because of their higher perspective and intelligence. In most Native American peoples, the Eagle clan was important and the Eagle was a messenger that brought with it hope and faith on the wings of spirit. The Eagle has a keen sense of detail and technical orientation. If the Eagle aspires for too much and seeks to attain it because of its superior abilities, this may lead to a fall from the heights. Eagles love freedom and should be aware of the risks of escapism, since they may escape problems simply by looking at them from a soaring perspective.

Owl (Cib)

Owls are people that embody the wisdom of the past and have unusual abilities of a psychic nature. It is not an easy day sign and may involve having a lot of karmic cleansing to do. On the surface, Owls may be jolly and humorous, but underneath they are very deep and serious. Usually Owl is referred to as Vulture, which points to the lazy side of this Southern sign. Owls are bent to introspection from which they derive much wisdom, and they often help their fellow humans with their psychic abilities. Their challenge in life is to find an ethical way of taking it easy.

Earth (Caban)

In contrast to what you might spontaneously think of the Earth sign, Earth people are very mental and they emphasize the value of thought processes. Earth is an Eastern energized sign, and there is a typical masculine side to Earth people that makes them want to control the world by understanding it intellectually. Earth people help disperse bad intentions, habits, and ideas. This turns them into good advisors and at their best, their thinking is in resonance with Mother Earth and serves the larger whole. Meticulous and intelligent, Earth people aspire to have a natural flow in all phases of life. Sometimes the sensitivity of Earth is shaken, which may lead to emotional disruptions; therefore, another name for Earth is Earthquake. Their challenge is not to let their intellect block them from being in the present.

Flint (Etznab)

Flint may be a fairly difficult day sign to be born into. Flints use their inner qualities to distinguish between truth and falseness. Often those born into this day sign are likened to the Archangel Michael, who by using his sword separates good from evil; other names for this day sign are Knife or Obsidian. In fact, many would probably find Flints opinionated and having too strong ideas about right and wrong. Flints are, however, very honest and want to serve by discerning the truth; they can spot hidden agendas in others from miles away. It was said that Flints could receive information about interpersonal problems or the evil plans of others by reflecting them in a mirror of obsidian glass. Their challenge is to find harmony and to use their powers of distinction as a gift rather than as a knife.

Rainstorm (Cauac)

Rainstorms are typically fun people to be with. They do not ever seem to age in spirit, and they keep curious about new things throughout their lives. Rainstorms are researchers who go on to study one thing after another, and in this way they will assemble a lot of knowledge in a lifetime. They have a great ability to both teach and learn. Yet, it is not always easy for them to synthesize the knowledge that they have gathered from so many sources and commit to a purpose in life. Forever young, they exist to enjoy the ecstasy of freedom. Their ongoing search for new experiences will lead them into big challenges and emotional storms, and many of them will experience that their entire life has been nothing but one long storm. They do have a sensitivity that will not always make it easy for them to face the storms of life. To see these difficulties as teachings will help Rainstorms find the greater wholeness they are looking for in life.

Light (Ahau)

Light is the day sign of completion, and this sign will tend to have consequences for those that are born into it. Light people are often romantic, enthusiastic visionaries with artistic abilities who are easily perceived as dreamers. It seems that because they have been born into this spiritual sign of completion, they find it difficult to understand that the world around them has not yet arrived at such a high state and instead is dominated by materialistic or greedy motives. Thus, in confronting "real life," they will often be perceived as unrealistic. A series of disappointments may then lead the Sun (Light) to shun responsibility and to not accept the necessary corrective measures. Nonetheless, Light people will retain a natural spirituality that is the birthright of this, the last sign in the uinal. For Light people, the challenge is to approach life realistically without compromising the great dreams they have.

YOUR COSMIC TONE

The Number you have found within parentheses in the Sacred Calendar is your Cosmic Tone. In certain respects your Cosmic Tone influences your Sacred Calendar energy of birth.

In the Sacred Calendar, it is possible to discern a pattern of alternating *light* and *dark* energies. This should not be interpreted as positive or negative. Rather, the divine process of creation is a wave movement of energies. In this way, activity (light fields) and passivity (dark fields) alternate. Your Cosmic Tone reflects a distinct step in an evolution from seed to mature fruit. In the days with the tone 1, a seed is planted and on days with the tone 13, a fruit matures. Every day in this process of growth represents a specific energy. For this reason, every day in the Sacred Calendar attains a unique energy.

Among the Aztecs—who used the same Sacred Calendar as the Maya, except with different names for some of the symbols—the thirteen numbers (or the Cosmic Tones) were each dominated by a special deity, which tells us something about the energies of these numbers. This deity gives rise to different characters among those that were born with the corresponding tones.

1. God of fire and time — *initiates*
2. God of the earth — *creates a reaction*
3. Goddess of water and childbirth — *activates*
4. God of warriors and the sun — *stabilizes*
5. Goddess of love and childbirth — *empowers*
6. God of death — *creates flow*
7. God of maize — *reveals*
8. God of rain — *harmonizes*
9. God of light — *creates movement forward*
10. God of darkness — *challenges*
11. Goddess of childbirth — *creates clarity*
12. God ruling before dawn — *creates understanding*
13. The supreme deity — *completes*

For further information about the Mayan Calendar, please visit the web pages:

www.mayanmajix.com
www.calleman.com
www.handclow2012.com

NOTES

CHAPTER 1: THE MAYAN CALENDAR

1. Kearsley, *Mayan Genesis,* 278.
2. Coe, *Breaking the Maya Code,* 123–35.
3. Argüelles, *Mayan Factor,* 45.
4. Hancock, *Fingerprints of the Gods;* Michell, *New View over Atlantis;* and Tompkins, *Mysteries of the Mexican Pyramids.*
5. Lockyer, *Dawn of Astronomy,* 243–48; Mehler, *From Light into Darkness,* 70; Sidharth, *Celestial Keys to the Vedas,* 60.
6. Jenkins, *Maya Cosmogenesis,* 299–311.
7. Freidel et al., *Maya Cosmos,* 59–122.
8. Ibid.
9. Mehler, *From Light into Darkness,* 70.
10. Jenkins, *Maya Cosmogenesis,* 31–35, 253–263, 273–79.
11. Ibid., 73–76.
12. Ibid., 256–63.
13. Ibid., 320.
14. Shearer, *Lord of the Dawn,* 184.
15. Argüelles, *Mayan Factor.*
16. Mann, *Shadow of a Star,* 84–86.
17. Argüelles, *Mayan Factor,* 146–48.
18. Ibid., 116–17.
19. Clow, *Catastrophobia.*

CHAPTER 2: ORGANIC TIME

1. Calleman, *Greatest Mystery of Our Time,* 97.
2. Ibid., 236–37.
3. Clow, *Catastrophobia.*
4. Calleman, *Greatest Mystery of Our Time,* 82.
5. Sidharth, *Celestial Keys to the Vedas,* 60.
6. Calleman, *Mayan Calendar,* 91.
7. Schick and Toth, *Making Silent Stones Speak,* 314.
8. Ibid., 143.
9. Ibid., 284.
10. Ibid., 293.

CHAPTER 3: THE GLOBAL MARITIME CIVILIZATION

1. Goodman, *Ecstasy, Ritual, and Alternate Reality,* 17.
2. Clow, *Catastrophobia;* Hancock, *Fingerprints of the Gods;* and Hapgood, *Ancient Sea Kings.*
3. Goodman, *Ecstasy, Ritual, and Alternate Reality,* 69.
4. Ibid., 70–87.
5. National Geographic News, "Did Island Tribes Use Ancient Lore to Evade Tsunami?," NationalGeographic.com, January 24, 2005.
6. Goodman, *Ecstasy, Ritual, and Alternate Reality,* 18.
7. Derek S. Allan, "An Unexplained Arctic Catastrophe. Part II. Some Unanswered Questions," *Chronology & Catastrophism Review,* 2005, 3–7.
8. National Geographic News, "Tribes Use Ancient Lore."
9. Allan and Delair, *Cataclysm!,* 250–54; and Clow, *Catastrophobia,* 39–40.
10. Jenkins, *Maya Cosmogenesis,* 116.
11. Hancock, *Underworld.*
12. Hapgood, *Ancient Sea Kings,* 188.
13. Settegast, *Plato Prehistorian,* 15–20.
14. Ibid., 23.
15. Ryan and Pitman, *Noah's Flood,* 188–201.
16. Blair, *Ring of Fire,* 57, 70–89.
17. Clow, *Catastrophobia,* 83–85.

18. Blair, *Ring of Fire,* 57; and Oppenheimer, *Eden in the East,* 147–55.
19. Allan and Delair, *Cataclysm!,* 40–42.
20. Schoch, *Voices of the Rocks,* 5–6, 33–56, 74–78, 242; and West, *Serpent in the Sky,* 198–209, 226–27.
21. Hancock, *Fingerprints of the Gods,* 357; Clow, *Catastrophobia,* 61–68; and Mehler, *Land of Osiris,* 186–88.
22. Clow, *Catastrophobia,* 61–63.
23. Hapgood, *Ancient Sea Kings,* 1–30, 32–33, 180–82; and Clow, *Catastrophobia,* 171–74, 177–81, 186.
24. Shick and Toth, *Making Silent Stones Speak,* 29.
25. Shick and Toth, *Making Silent Stones Speak,* 286–93; and Collins, *Ashes of Angels,* 247.
26. Shick and Toth, *Making Silent Stones Speak,* 286–301.
27. Goodman, *Ecstasy, Ritual, and Alternate Reality,* 86.
28. Calleman, *Mayan Calendar,* 113.
29. Ibid., 115.
30. Ibid., 115.
31. Ibid., 118.
32. Mavor and Dix, *Manitou.*
33. Ibid., 103–17.
34. Dunn, *Giza Power Plant.*
35. Ibid., 138, 219.
36. Ibid., 109–19, 234.
37. Ibid., 114–15.
38. Clow, *Catastrophobia,* 87; and Settegast, *Plato Prehistorian,* 24–26, 106–11.
39. Settegast, *Plato Prehistorian,* 27.
40. Clottes and Courtin, *Cave Beneath the Sea,* 34–35.
41. Clow, *Catastrophobia,* 87–94.
42. Devereux, *Stoneage Soundtracks,* 110–15; and Clow, *Alchemy of Nine Dimensions,* 111–14.
43. 1989 Mayan Initiatic Journeys arranged by Maya Mysteries School, Apdo. Postal 7-014, Merida 7, Yucatan, Mexico.
44. Devereux, *Stoneage Soundtracks,* 76–89.
45. Clow, *Alchemy of Nine Dimensions,* 115–18.
46. Strong, "Carnac, Stones for the Living," 62–79.

47. Clow, *Alchemy of Nine Dimensions,* 117.
48. John Beaulieu, BioSonic Enterprises, P.O. Box 487, High Falls, NY 12440. On the Web at www.BioSonicEnterprises.com.
49. Cuyamungue Institute, P.O. Box 2202, Westerville, OH 43086. On the Web at www.CuyamungueInstitute.com.
50. Gore, *Ecstatic Body Postures,* ix.
51. Gore, *Ecstatic Body Postures,* ix; and Goodman and Nauwald, *Ecstatic Trance,* 18–19.
52. Goodman, *Ecstasy, Ritual, and Alternate Reality,* 39.
53. Ibid.
54. Gore, *Ecstatic Body Postures,* 173–78, 202–8, 241–44.

CHAPTER 4: ENTERING THE MILKY WAY GALAXY

1. Calleman, *Mayan Calendar,* 52.
2. Swimme, *Hidden Heart of the Cosmos,* 80–81.
3. Glanz, "Cosmic Boost."
4. Calleman, *Mayan Calendar,* 117–18.
5. Emoto, *Hidden Messages in Water.*
6. Calleman, *Mayan Calendar,* 107.
7. Gillette, *Shaman's Secret,* 47.
8. Jenkins, *Galactic Alignment,* 249.
9. Clow, *Alchemy of Nine Dimensions,* 144.
10. "The Strange Case of Earth's New Girth," *Discover,* January 2003, 52; and Clow, *Alchemy of Nine Dimensions,* 150.
11. "Earth's New Girth," 52; and Clow, *Alchemy of Nine Dimensions,* 150.
12. Clow, *Alchemy of Nine Dimensions,* 149.
13. Glanz, "Cosmic Boost"; and Clow, *Alchemy of Nine Dimensions,* 149.
14. Clow, *Alchemy of Nine Dimensions,* 149.
15. Ibid.
16. Bentov, *Stalking the Wild Pendulum,* 134–39.
17. Ibid., 137.
18. Clow, *Alchemy of Nine Dimensions,* 145–48.
19. Ibid., 144.

20. Overbye, "Other Dimensions?"
21. Sidharth, *Celestial Keys to the Vedas.*
22. Ibid., 60.
23. Ibid., 25.
24. Yukteswar, *Holy Science,* x–xxi.
25. Sidharth, *Celestial Keys to the Vedas,* 34–38, 107.
26. Kearsley, *Mayan Genesis.*
27. Jenkins, *Galactic Alignment,* 238.
28. Ibid.
29. Ibid.
30. Ibid.
31. LaViolette, *Earth Under Fire,* 306–9; and Allan and Delair, *Cataclysm!,* 209–10.
32. Jenkins, *Galactic Alignment,* 238–48; and Argüelles, *Earth Ascending,* 15–26.

CHAPTER 5: THE WORLD TREE

1. Calleman, *Greatest Mystery of Our Time,* 35–37.
2. Tedlock, *Popol Vuh.*
3. Gillette, *Shaman's Secret,* 34.
4. Ibid., 30–31.
5. Calleman, *Greatest Mystery of Our Time,* 37.
6. On the Web with Matthew Fox at www.matthewfox.org.
7. Calleman, *Mayan Calendar,* 36.
8. Argüelles, *Mayan Factor,* 109–30.
9. Calleman, *Greatest Mystery of Our Time,* 37.
10. Clow, *Mind Chronicles,* 356–63.
11. Argüelles, *Earth Ascending,* 15–26.
12. Calleman, *Mayan Calendar,* 50–51.
13. Ibid., 51.
14. Van Andel, *New Views,* 135.
15. Calleman, *Mayan Calendar,* 43.
16. Calleman, *Greatest Mystery of Our Time,* 39–46.
17. Calleman, *Greatest Mystery of Our Time,* 39–46; and Calleman, *Mayan Calendar,* 36–45.

18. Calleman, *Greatest Mystery of Our Time,* 44.
19. Calleman, *Mayan Calendar,* 41.
20. Ibid., 42.
21. Calleman, *Greatest Mystery of Our Time,* 46.
22. Ibid., 47.
23. Clow, *Alchemy of Nine Dimensions,* 9.
24. Calleman, *Mayan Calendar,* 54–58.
25. Ibid., 56.
26. Clow, *Alchemy of Nine Dimensions,* 111–12.
27. Calleman, *Mayan Calendar,* 59.
28. Ibid., 58.
29. Ibid.
30. Ibid., 59.
31. Ibid., 60.
32. Ibid., 62.
33. Sidharth, *Celestial Keys to the Vedas,* 35.
34. Goodman and Nauwald, *Ecstatic Trance,* 23–25.

CHAPTER 6: THE GALACTIC UNDERWORLD AND TIME ACCELERATION

1. Garrison, *America as Empire.*
2. On the Web at www.commondreams.org/views04/0225-05.htm.
3. Ron Suskind, *The Price of Loyalty.*
4. David Icke, *Alice in Wonderland and the World Trade Center Disaster;* John Kaminski, *America's Autopsy Report;* Michael C. Ruppert, *Crossing the Rubicon;* David Ray Griffin, *The New Pearl Harbor.* On the Web, see www.globaloutlook.ca and www.journalof911studies.com.
5. Ruppert, *Crossing the Rubicon,* 22–150.
6. Calleman, *Mayan Calendar,* 65.
7. Ibid., 149–50.
8. Ibid., 151.
9. Ibid., 151.
10. Bob Drogin, "Through the Looking Glass."
11. Calleman, *Mayan Calendar,* 148.
12. Brown, *Da Vinci Code.*
13. Hancock, *Fingerprints of the Gods.*

14. David Stipp, "Climate Collapse."
15. Clow, *Alchemy of Nine Dimensions*.
16. On the Web at www.wiseawakening.com.
17. Harvey, *Sun at Midnight;* and www.gracecathedrale.org/archives.
18. Ibid.
19. Calleman, *Mayan Calendar,* 142.
20. Ibid., 161–62.

CHAPTER 7: ENLIGHTENMENT AND PROPHECY THROUGH 2011

1. Calleman, *Mayan Calendar,* xvii.
2. Ibid., 120–36.
3. Ibid., 139.
4. Ibid., 139.
5. Brown, *Angels and Demons*.
6. Calleman, *Mayan Calendar,* 257.
7. Ibid., 177–78.
8. Ibid., 177.
9. Ibid., xix.
10. Urquhart, *The Pope's Armada*.
11. Kunstler, *The Long Emergency,* 6.
12. Ibid., 5–6.
13. Wente, "Watch Out! More Health Care Can Really Make You Sick."
14. Calleman, *Mayan Calendar,* 157.
15. Ibid., 145.
16. Ibid., 161.
17. Webre, *Exopolitics*.
18. Ibid., 12–13, 16.
19. Ibid., 6.
20. Calleman, *Mayan Calendar,* 158.
21. Dick, *Biological Universe,* 476.
22. Webre, *Exopolitics,* 11.
23. Ibid., 11–12.
24. Ibid., 13.
25. Rudgley, *Lost Civilizations,* 100.
26. Webre, *Exopolitics,* 14.

27. Ibid., 16.
28. Ibid., 17.
29. Ibid., 33.
30. Ibid., 18.

CHAPTER 8: CHRIST AND THE COSMOS

1. LaViolette, *Earth Under Fire.*
2. LaViolette, *Message of the Pulsars,* 58.
3. Allan and Delair, *Cataclysm!,* 209.
4. LaViolette, *Message of the Pulsars,* 69–70; and Allan and Delair, *Cataclysm!,* 209.
5. On the Web at www.crawford2000.co.uk/planetchange1.htm.
6. LaViolette, *Message of the Pulsars,* 143–66.
7. Clow, *Mind Chronicles.*
8. LaViolette, *Message of the Pulsars.*
9. LaViolette, *Message of the Pulsars,* 1–4; and on the Web at Wikipedia, "Pulsars," en.wikipedia.org/wiki/Pulsar.
10. LaViolette, *Genesis of the Cosmos,* 181, 222.
11. Ibid., 181.
12. Ibid., 218–21, 288–95.
13. LaViolette, *Message of the Pulsars,* 16.
14. Ibid., 20–27.
15. Ibid., 27.
16. Ibid., 28–29.
17. Ibid., 29.
18. Ibid., 30–32.
19. Ibid., 32.
20. Ibid., 33–43.
21. Ibid., 38.
22. Ibid., 40.
23. Ibid., 98–113.
24. Ibid., 109.
25. Ibid., 143–166.
26. Ibid., 109.
27. Ibid., 143–47.

28. Wansbrough, *New Jerusalem Bible,* Book of Daniel, 4:10–23.
29. Collins, *Ashes of Angels,* 247–48; and Solecki, Shanidar, *The First Flower People.*
30. Collins, *Ashes of Angels,* 248–51.
31. Ibid., 250
32. Ibid., 36, 251.
33. O'Brien, *Genius of the Few,* 122–27.
34. Ibid., 72.
35. Knight and Lomas, *Uriel's Machine.*
36. Ibid., 397.
37. Ibid., 231, 289.
38. Ibid., 287.
39. Ibid., 231.
40. Laurence, *Book of Enoch,* 5–8; and Lawton, *Genesis Unveiled,* 47–49.
41. Knight and Lomas, *Uriel's Machine,* 344.
42. Ibid., 95–100, 220–235, 339–347, 361–68.
43. Gaffney, *Gnostic Secrets,* 20–31.
44. Atwater, *Indigo Children.*
45. Knight and Lomas, *Uriel's Machine,* 324.
46. Ibid., 93–95.
47. Shanks and Witherington, *Brother of Jesus,* 93–125; and Butz, *Brother of Jesus,* 50–103.
48. Knight and Lomas, *Uriel's Machine,* 326.
49. Ibid., 326.
50. Ibid., 95.
51. Ibid., 326.
52. Strachan, *Jesus the Master Builder.*
53. Knight and Lomas, *Uriel's Machine,* 327.
54. Strachan, *Jesus the Master Builder,* 10–11.
55. Ibid., 83, 85, 227.
56. Ibid., 88.
57. Ibid., 118–125.
58. Ibid., 119.
59. Hassnain, *Search for the Historical Jesus,* 168–69; Baigent, *The Jesus Papers,* 17, 121.

60. Strachan, *Jesus the Master Builder,* 120.
61. Lawton, *Genesis Unveiled.*
62. Ibid., 221–235.
63. Ibid., 46–47.
64. Clow, *Catastrophobia,* 138–40, 162–63, 171.
65. Collins, *Ashes of Angels,* 14–16.
66. Lawrence, *Book of Enoch,* 5–8; and Collins, *Ashes of Angels,* 230–39.
67. Lawton, *Genesis Unveiled,* 50.
68. Ibid.
69. Ibid.
70. Ibid.
71. Shick and Toth, *Making Silent Stones Speak,* 81–82, 261–62.
72. Lawton, *Genesis Unveiled,* 108.
73. Ibid., 108.
74. Tedlock, *Popol Vuh,* 79–86, 163–67.
75. Ibid., 85–86.
76. Lawton, *Genesis Unveiled,* 111.
77. Ibid.
78. Ibid., 150.
79. Ibid., 150.
80. Tedlock, *Popol Vuh,* 163–66.
81. Ibid., 166–67.
82. Lawton, *Genesis Unveiled,* 115.
83. Cory, *Ancient Fragments,* 20.
84. Lawton, *Genesis Unveiled,* 121.

APPENDIX A: REFLECTIONS ON EARTH'S TILTING AXIS

1. J. B. Delair, "Planet in Crisis," *Chronology and Catastrophism Review* (1997), 4–11. Only sections of the article are excerpted.
2. Marshack, *Roots of Civilization,* 9–16.
3. Ibid.
4. Rudgley, *Lost Civilizations,* 102.
5. Ibid., 102–4.
6. Ibid., 104.

7. Brennan, *The Stars and the Stones: Ancient Art and Astronomy in Ireland.*
8. Temple, *The Crystal Sun.*
9. Ellis, *Thoth: Architect of the Universe,* 104–31.
10. Knight and Lomas, *Uriel's Machine,* 152–82.
11. Ibid., 213–32.
12. Lockyer, *Dawn of Astronomy,* 108.
13. Temple, *Crystal Sun,* 412–14.

BIBLIOGRAPHY

Allan, D. S., and J. B. Delair. *Cataclysm!* Santa Fe: Bear & Company, 1997.

Argüelles, José. *The Mayan Factor.* Santa Fe: Bear & Company, 1987.

———. *Earth Ascending.* Santa Fe: Bear & Company, 1996.

Atwater, P. M. H. *Beyond the Indigo Children.* Rochester, Vt.: Bear & Company, 2005.

Baigent, Michel. *The Jesus Papers.* San Francisco: HarperCollins, 2006.

Bentov, Itzhak. *Stalking the Wild Pendulum.* Rochester, Vt.: Destiny Books, 1988.

Blair, Lawrence. *Ring of Fire.* New York: Bantam Books, 1988.

Brennan, Martin. *The Stars and the Stones: Ancient Art and Astronomy in Ireland.* London: Thames and Hudson, 1985.

Brown, Dan. *Angels and Demons.* New York: Pocket Books, 2000.

———. *The Da Vinci Code.* New York: Doubleday, 2003.

Butz, Jeffrey J. *The Brother of Jesus.* Rochester, Vt.: Inner Traditions, 2005.

Calleman, Carl Johan. *Solving the Greatest Mystery of Our Time.* Coral Springs, Fla.: Garev Publishing International, 2001.

———. *The Mayan Calendar and the Transformation of Consciousness.* Rochester, Vt.: Bear & Company, 2004.

Clottes, Jean, and Jean Courtin. *Cave Beneath the Sea.* New York: Harry Abrams, 1996.

Clow, Barbara Hand. *Catastrophobia.* Rochester, Vt.: Bear & Company, 2001.

———. *The Mind Chronicles*. Rochester, Vt.: Bear & Company, 2007.

Clow, Barbara Hand, with Gerry Clow. *Alchemy of Nine Dimensions*. Charlottesville, Va.: Hampton Roads, 2004.

Coe, Michael D. *Breaking the Maya Code*. New York: Thames & Hudson, 1993.

Collins, Andrew. *From the Ashes of Angels*. London: Penguin Books, 1996.

Cory, Isaac Preston. *Ancient Fragments*. Savage, Minn.: Wizards Bookshelf, 1975.

Devereux, Paul. *Stoneage Soundtracks*. London: Vega, 2001.

Dick, Steven J. *The Biological Universe*. New York: Cambridge University Press, 1996.

Drogin, Bob. "Through the Looking Glass into the Mind of Saddam." *Austin American Statesman*, October 15, 2004, sec. A, 25–26.

Dunn, Christopher. *The Giza Power Plant*. Santa Fe: Bear & Company, 1998.

Ellis, Ralph. *Thoth: Architect of the Universe*. Dorset, U.K.: Edfu Books, 1997.

Emoto, Masuro. *The Hidden Messages in Water*. Hillsboro, Ore.: Beyond Words Publishing, 2004.

Freidel, David, Linda Schele, and Joy Parker. *Maya Cosmos*. New York: William and Morrow, 1993.

Gaffney, Mark H. *Gnostic Secrets of the Naassenes*. Rochester, Vt.: Inner Traditions, 2004.

Garrison, Jim. *America as Empire*. San Francisco: Berrett-Koehler Publishers, 2004.

Gillette, Douglas. *The Shaman's Secret*. New York: Bantam Books, 1997.

Glanz, James. "Theorists Ponder a Cosmic Boost from Far, Far Away." *New York Times*, February 15, 2000.

Goodman, Felicitas D. *Where the Spirits Ride the Wind*. Bloomington, Ind.: Indiana University Press, 1990.

———. *Ecstasy, Ritual, and Alternate Reality*. Bloomington, Ind.: Indiana University Press, 1992.

Goodman, Felicitas, and Nana Nauwald. *Ecstatic Trance*. Havalte, Holland: Binkey Kok Publications, 2003.

Gore, Belinda. *Ecstatic Body Postures*. Santa Fe: Bear & Company, 1995.

Griffin, David Ray. *The New Pearl Harbor.* Northampton, Mass.: Olive Branch Press, 2004.

Hancock, Graham. *Fingerprints of the Gods.* New York: Crown Publishing, 1995.

———. *Underworld.* New York: Crown Publishers, 2002.

Hapgood, Charles. *Maps of the Ancient Sea Kings.* London: Turnstone Books, 1979.

Hassnain, Fida. *A Search for the Historical Jesus.* Bath, U.K.: Gateway Books, 1944.

Harvey, Andrew. *The Sun at Midnight: A Memoir of the Dark Night.* New York: Tarcher, 2002.

Icke, David. *Alice in Wonderland and the World Trade Center Disaster.* Wildwood, Mo.: Bridge of Love Publications, 2002.

Jenkins, John Major. *Galactic Alignment.* Rochester, Vt.: Bear & Company, 2002.

———. *Maya Cosmogenesis 2012.* Santa Fe: Bear & Company, 1998.

Kaminsky, John. *America's Autopsy Report.* Tempe, Ariz.: Dandelion Books, 2003.

Kearsley, Graeme R. *Mayan Genesis.* London: Yelsraek Publishing, 2001.

Knight, Christopher, and Robert Lomas. *Uriel's Machine: The Prehistoric Technology That Survived the Flood.* Boston: Element Books, 2000.

Kunstler, James Howard. *The Long Emergency.* New York: Atlantic Monthly Press, 2005.

Laurence, Richard, translator. *The Book of Enoch.* San Diego: Wizards Bookshelf, 1973.

LaViolette, Paul A. *Decoding the Message of the Pulsars.* Rochester, Vt.: Bear & Company, 2006.

———. *Earth Under Fire.* Rochester, Vt.: Bear & Company, 2005.

———. *Genesis of the Cosmos.* Rochester, Vt.: Bear & Company, 2004.

Lawton, Ian. *Genesis Unveiled.* London: Virgin Books, 2003.

Lockyer, J. Norman. *The Dawn of Astronomy.* Kila, Mont.: Kessinger Publishing, 1997.

Mann, Alfred K. *Shadow of a Star.* New York: W. H. Freeman and Company, 1997.

Marshack, Alexander. *The Roots of Civilization.* New York: McGraw-Hill, 1967.

Mavor, James W., and Byron E. Dix. *Manitou*. Rochester, Vt.: Inner Traditions, 1989.

Mehler, Stephen S. *From Light into Darkness*. Kempton, Ill.: Adventures Unlimited, 2005.

———. *The Land of Osiris*. Kempton, Ill.: Adventures Unlimited, 2001.

Michell, John. *A New View over Atlantis*. San Francisco: Harper & Row, 1983.

O'Brien, Christian. *The Genius of the Few*. Wellingborough, Northamptonshire, U.K.: Turnstone, 1985.

Oppenheimer, Stephen. *Eden in the East*. London: Weidenfeld and Nicolson, 1998.

Overbye, Dennis. "Other Dimensions? She's in Pursuit." *New York Times,* September 30, 2003.

Rudgley, Richard. *The Lost Civilizations of the Stone Age*. New York: Free Press, 1999.

Ruppert, Michael C. *Crossing the Rubicon*. Gabriola Island, British Columbia: New Society Publishers, 2004.

Ryan, William, and Walter Pitman. *Noah's Flood*. New York: Simon and Schuster, 1998.

Schick, Kathy D., and Nicholas Toth. *Making Silent Stones Speak*. New York: Simon and Schuster, 1993.

Schoch, Robert M. *Voices of the Rocks*. New York: Harmony House, 1999.

Settegast, Mary. *Plato Prehistorian*. Hudson, N.Y.: Lindesfarne Press, 1990.

Shanks, Hershel, and Ben Witherington III. *The Brother of Jesus*. San Francisco: HarperCollins, 2003.

Shearer, Tony. *Lord of the Dawn: Quetzalcoatl*. Happy Camp, Calif.: Naturegraph, 1971.

Sidharth, B. G. *The Celestial Keys to the Vedas*. Rochester, Vt.: Inner Traditions, 1999.

Stipp, David. "Climate Collapse." *Fortune,* January 26, 2004, 14–22.

Strachan, Gordon. *Jesus the Master Builder*. Edinburgh: Floris Books, 1999.

Strong, Roslyn. "Carnac, Stones for the Living: A Megalithic Seismograph?" *NEARA Journal* 35 (no. 2, Winter 2001) 62–79.

Suskind, Ron. *The Price of Loyalty*. New York: Simon and Schuster, 2004.

Swimme, Brian. *The Hidden Heart of the Cosmos*. Maryknoll, N.Y.: Orbis, 1996.

Tedlock, Dennis. *Popol Vuh*. New York: Simon and Schuster, 1986.

Temple, Robert. *The Crystal Sun*. London: Century Books, 2000.

Tompkins, Peter. *Mysteries of the Mexican Pyramids*. San Francisco: Harper & Row, 1976.

Urquart, Gordon. *The Pope's Armada*. Amherst, N.Y.: Prometheus Books, 1999.

Van Andel, Tjerd. *New Views on an Old Planet*. New York: Cambridge University Press, 1994.

Wansbrough, Henry, editor. *The New Jerusalem Bible*. New York: Doubleday, 1985.

Webre, Alfred Lambremont. *Exopolitics*. Vancouver, British Columbia: Universebooks, 2005.

Wente, Margaret. "Watch Out! More Health Care Can Make You Sick." *Globe and Mail*, May 25, 2006.

West, John Anthony. *Serpent in the Sky*. New York: Harper & Row, 1979.

Yukteswar, Swami Sri. *The Holy Science*. Los Angeles: Self-Realization Fellowship, 1977.

INDEX

BOOKS OF RELATED INTEREST

The Pleiadian Agenda
A New Cosmology for the Age of Light
by Barbara Hand Clow

Catastrophobia
The Truth Behind Earth Changes
by Barbara Hand Clow

The Mayan Calendar and the Transformation of Consciousness
by Carl Johan Calleman, Ph.D.

Maya Cosmogenesis 2012
The True Meaning of the Maya Calendar End-Date
by John Major Jenkins

Galactic Alignment
The Transformation of Consciousness According to Mayan, Egyptian, and Vedic Traditions
by John Major Jenkins

The Mayan Oracle
Return Path to the Stars
by Ariel Spilsbury and Michael Bryner
Illustrated by Donna Kiddie

How to Practice Mayan Astrology
The Tzolkin Calendar and Your Life Path
by Bruce Scofield and Barry C. Orr

Time and the Technosphere
The Law of Time in Human Affairs
by José Argüelles